Waste as a Resource

Series Editor

Cristina Trois, Civil Engineering, Stellenbosch University, Stellenbosch, South Africa

The Waste as a Resource series explores state-of-the-art knowledge and innovation in waste management strategies focused on approaches and technologies that ensure maximum value recovery from waste through materials recycling and energy production. The series aims to navigate the transition toward a circular economy by presenting techniques and strategies for waste minimization, effective waste diversion from landfills, and, most importantly, exploring the possible pathways for treating waste as a resource. The series has a global scope, focusing on developing countries and emerging economies. This will assist in filling the gap in current publications that need to adequately explore valorization technologies that are appropriate for and adaptable in low-income countries or informal contexts.

Alexander Meneses-Jácome
Ulrike Schmid-Staiger • Tatiana C. Guarín C.
Editors

The Circular Bioeconomy in Industry

Strategies for Sustainable Resource Recovery, Environmental Balance, and Ecosystem Protection

Springer

Editors
Alexander Meneses-Jácome
Circular Bio-Economy Research Center
(ε-BiO) and Environmental Biotechnology
Research Group
Universidad Autónoma de
Bucaramanga - UNAB
Bucaramanga, Santander, Colombia

Ulrike Schmid-Staiger
Fraunhofer Institute for Interfacial
Engineering and Biotechnology IGB
Stuttgart, Baden-Württemberg, Germany

Tatiana C. Guarín C.
Grupo de Investigación en Recursos
Hídricos y Saneamiento Ambiental - GPH
Escuela de Ingeniería Civil
Universidad Industrial de Santander – UIS
Bucaramanga, Santander, Colombia

ISSN 2731-8222 ISSN 2731-8230 (electronic)
Waste as a Resource
ISBN 978-3-031-84358-7 ISBN 978-3-031-84359-4 (eBook)
https://doi.org/10.1007/978-3-031-84359-4

This Springer imprint is published by the registered company Springer Nature Switzerland AG
The registered company address is: Gewerbestrasse 11, 6330 Cham, Switzerland

Preface

For more than two decades, climate action has prevailed by far as the main need and road map for the search for a more sustainable and rational socioeconomic development. However, the extreme focus on this aspect has led to a systematic neglect of other environmental impacts and sustainability aspects, whose effects are more sensitive at the territorial scale. This situation has triggered a stronger motivation to change the current economic model, but with a greater emphasis on the efficient use of natural resources and their conservation.

It is at this crossroads that the "circular economy" emerges, a "pro-sustainability" economic model that focuses on the functionalization of production systems as "regenerative systems," i.e., reducing the input of raw materials and the output of waste, emissions, and energy, in order to curb the consumption of resources and reduce intervention in ecosystems and natural cycles. This approach adopts design methods focused on facilitating the recovery of materials from products when they reach their end of life, allowing their permanence in the economic system to be extended or their return to nature in the best possible conditions of energy efficiency and environmental protection. So, possibly, this tangible and methodical move toward sustainability explains the shift of environmental sustainability policies in different contexts toward "circularity" models, a notion that is perceived as more pragmatic, measurable, or verifiable than the notion of sustainability itself and which, therefore, seems capable of controlling or mitigating the traumas inherent to technological transitions due to or associated with changes in economic models.

Complementing the intentionality of the principles of the circular economy appears "bioeconomy," which promotes the rationalization of the use of nonrenewable resources, turning to a more widespread use of "more renewable resources," i.e., ecosystems and biological resources, which, despite this connotation, also depend, although to a lesser extent, on ores, but which have progressively lost their inherent capacity for regeneration and reduced their contribution to the regulation of the climate. Thus, the Organisation for Economic Co-operation and Development (OECD) defines the bioeconomy as the use of bio-based knowledge to obtain sustainable, eco-efficient, and competitive products from the rational exploitation of terrestrial and aquatic bio-resources, as well as "agro-wastes" and forest residues for food, industrial, and power production.

Precisely, the availability of residual biomass in almost all regions of the planet and the circulation of nutrients used in agriculture, both in these materials and in wastewater from agricultural and agro-industrial processes, are an opportunity and a stimulus for bioeconomy models to find a true point of intersection between the bioeconomy and the circular economy, which has been called the "circular bioeconomy." A very narrow domain that privileges the production of bio-products of high commercial value for different sectors and markets (agriculture, livestock, energy, processed foods, beverages, chemicals, health, pharmaceuticals, cosmetics, etc.) and the search for alternative products to those derived from petrochemicals, leaving a small section for energy recovery through the production of second- and third-generation bioenergy.

For Colombia, an emerging country, the transition toward these models of sustainable economy is perceived as a route that has not been simple but has been defined since the middle of the last decade, with the implementation of the Green Growth Mission, which has identified the bioeconomy, biodiversity, and biotechnology, as a priority core for the country's development, which was recognized and enhanced by the "Mission of Wise Ones" of 2019, generating the input that places this transformation within the "National Development Plan 2022–2026" within the actions called "productive economy through reindustrialization and bioeconomy." In this transition, the initiative to leverage "pro" economic models that are sustainable and more harmonious with the environment is clearly revealed, and in this context, there are countless opportunities for agri-food production chains in Colombia, represented by poultry, palm, cocoa, organic coffees, among others, which can generate new products from the former "waste." This evolution must necessarily be inclusive and bring together actors from the quadruple helix (government, business, society, university-science) to reach agreements on the appropriate allocation of science, technology, and innovation funds to support the development of the bioeconomy and its technological base.

It is along these lines that international cooperation is necessary to accelerate the implementation of robust and adaptable technologies in countries and regions with lower technical and economic capacities and to move toward an effectively sustainable bioeconomy. In this sense, Colombia has taken Germany, a leading country in cutting-edge biotechnology, as a model to follow in the development of its own bioeconomy model, and in 2020, through the "International Bioeconomy" program, with joint funding from the Ministry of Science, Technology and Innovation of Colombia (Minciencias) and the Federal Ministry of Education and Research (BMBF), funding was granted to four model projects, which are the origin and central axis of the International Workshop on Circular Bioeconomy, held in Bucaramanga, Colombia, whose proceedings and main results are contained in this book of summaries.

With this starting point and the convening of the leaders of the four projects in Colombia and Germany, representatives of academia, research, government agencies, industry, and civil society have been brought together to share a series of experiences and projects, which have resulted in an exuberant exhibition of capabilities and opportunities for progress in the bioeconomy and the circular economy, which we invite you to learn here.

Bucaramanga, Santander, Colombia — Alexander Meneses-Jácome
Stuttgart, Baden-Württemberg, Germany — Ulrike Schmid-Staiger
Bucaramanga, Santander, Colombia — Tatiana C. Guarín C.

Contents

Part I
Biotechnology, Bioproducts & Bioenergy

Recent Developments in Biorefinery with Special Focus on Industrial Biotechnology

Thomas Hahn and Susanne Zibek

Highlights

- An overview of the latest and recent research activities of the Bioprocess Development Group at Fraunhofer IGB is provided. It includes a comprehensive summary of the results concerning plant oil utilisation for long-chain dicarboxylic acid, epoxide and biosurfactant production, the production of microbial monomers (malic acid) and polymers (polyhydroxyalkanoates), and the direct extraction of chitin and chitosan from agricultural waste streams.
- A number of conclusions and assessments have been drawn that identify the future potential and the necessary optimisations to exploit the potential of these novel bioproducts.

1 Introduction

Industrial biotechnology plays a key role in creating a sustainable bioeconomy and moving away from fossil raw materials. As part of the biorefinery concept, it enables the use of biomass and organic waste materials for the production of chemicals, materials and energy, thereby reducing dependence on oil and contributing to a reduction in greenhouse gas emissions. Thus, industrial biotechnology is a diversified cross-sectional technology combining different areas such as molecular biology, bioanalytics, bioprocess engineering and biocatalysis. The latter is the core of

T. Hahn (✉) · S. Zibek
Group Bioprocess development, Innovation field Industrial Biotechnology, Fraunhofer IGB, Stuttgart, Germany
e-mail: thomas.hahn@igb.fraunhofer.de

A. Meneses-Jácome et al. (eds.), *The Circular Bioeconomy in Industry*, Waste as a Resource, https://doi.org/10.1007/978-3-031-84359-4_1

this discipline supported by the others. With the help of biotechnological processes, microorganisms, such as bacteria and fungi, or enzymes can be used to catalyse complex biological processes. These processes often take place at lower temperatures and pressures than conventional chemical processes, which reduces energy consumption and the formation of side products. The selection of suitable microbial tools thus enables the tailor-made production of a desired product. However, the greatest strength of the concept is the exploitation of a broad range of substrates, respectively, matrices for the production: from highly purified monosaccharides over partially purified biomass to pure waste streams. In this sense, the combination with chemical or other pretreatment processes to fractionate the biomass is not contradictory, but rather complementary, allowing industrial biotechnology to fully exploit its potential. Biomass fractionation or pretreatment is often essential to obtain a substrate that is more accessible to biological processes such as those carried out in industrial biotechnology (see Fig. 1). The fractions could then be used as

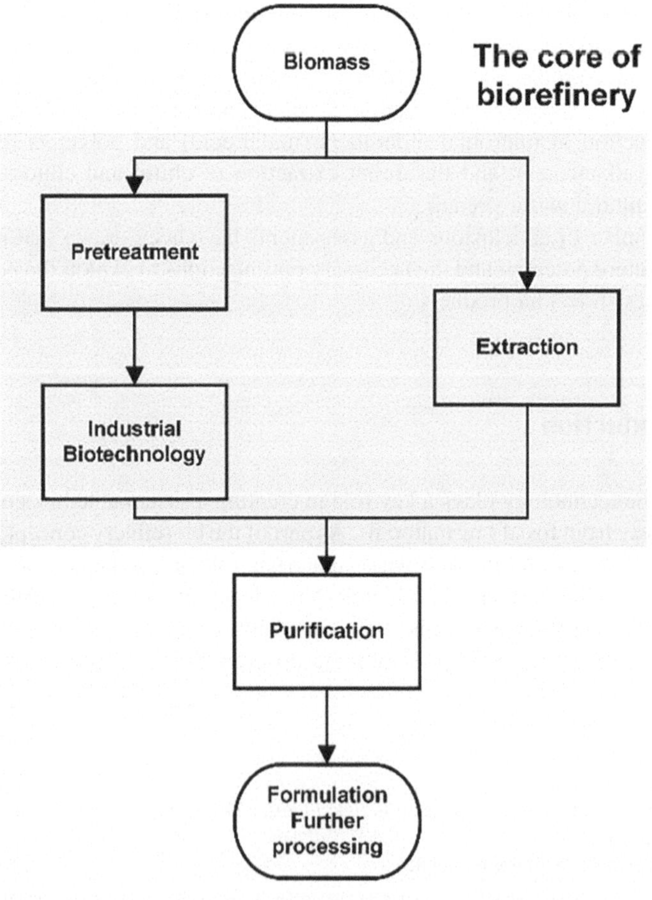

Fig. 1 Embedding of industrial biotechnology and different unit operations in biorefinery

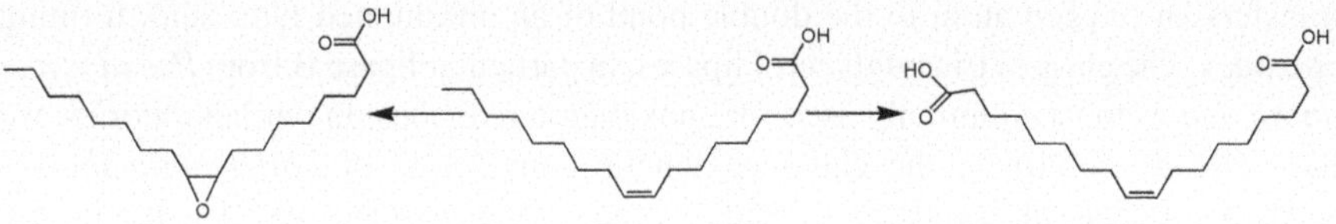

Fig. 2 Epoxidised oleic acid (left) and 1,18-octadecenedioic acid (right) by conversion of oleic acid using our chemoenzymatic approach or whole-cell catalysis

a substrate for fermentation, but also for enzymatic conversion. Another way to use the biomass is to extract a desired compound or to remove all undesired compounds, leaving the target product. Both the extracted compounds and the products derived from biotechnological processes are subjected to purification and finally to formulation or further processing.

In this context, we present a selection of promising biorefinery approaches that have recently been or are currently being pursued in the Bioprocess Development Group at the Fraunhofer IGB. These include the use of vegetable oils for the production of biosurfactants, epoxidised vegetable oils and long-chain dicarboxylic acids (lcDCAs). Another part of the review focuses on the microbial production of polyhydroxyalkanoates with carbohydrates and fatty acids, which demonstrates the potential to adapt or fine-tune the properties of a polymer by selecting suitable microorganisms and substrates. As an example of a platform chemical, current activities in L-malic acid fermentation—to be used as a monomer building block for polymalate production—are described in the next section. Another focus of the group is the extraction of chitin from waste streams. Our aim is to optimise the chitin purification process and subsequent conversion to chitosan, its acid-soluble derivative. This is described in the last section of this review.

Vegetable oils are widely used renewable resources in the chemical industry and are often modified to increase their reactivity. Here, we present several methods for the valorisation and utilisation of vegetable oils. Figure 2 shows the direct fatty acid and lipid functionalisation strategies.

2 Chemoenzymatic Epoxidation of Vegetable Oils

One example of such modification is the production of epoxides from vegetable oils. Epoxides derived from vegetable oils are suitable for use in adhesives, plasticisers or coatings or, for example, as stabilisers in chlorinated rubber. The epoxidation of vegetable oils, especially soybean oil as the best-known representative, is already established as a chemical process on an industrial scale. Chemoenzymatic epoxidation is a very promising alternative approach to overcome the problems associated with the chemical epoxidation process, such as unwanted ring opening or by-product formation. In the chemoenzymatic epoxidation process, lipases are used as biocatalysts to produce the required fatty peracid with hydrogen peroxide. The fatty peracid in turn transfers an oxygen atom to the double bond of an unsaturated fatty acid, forming epoxides. The class of triacylglycerol lipases, in particular lipase B from *Pseudozyma antarctica*, is most commonly used for epoxidation reactions. In our laboratories, we have already confirmed the chemoenzymatic epoxidation of *Lallemantia iberica* seed oil with epoxide yields >90% and oxirane oxygen contents >9.5% using immobilised lipase [1]. In addition, we carried out a scale-up of the reaction to the 100 L scale and used the same enzyme batch repeatedly in four consecutive cycles without losing an epoxidation yield of 70%. Other recent studies have reported the chemoenzymatic exploitation of waste oils, fatty acid derivatives or even other unsaturated substrates not structurally related to fatty acids. A further focus of the investigations is the optimisation of reaction conditions and systems. However, we strongly believe that further investigations should consider enzyme engineering approaches to increase the stability of the lipase in an oxidative environment as well as investigating further (bio-based) resins for immobilisation of the enzyme.

3 Long-Chain Dicarboxylic Acids from Fatty Acids

The previous method exploits the presence of unsaturated bonds in the vegetable oils/fatty acids. In contrast, microbial functionalisation to obtain long-chain dicarboxylic acids (lcDCAs) takes place on the non-reactive terminal chain of fatty acids, irrespective of the degree of saturation. LcDCAs and their esters are chemical intermediates widely used in the production of perfumes, polymers, adhesives and macrolide antibiotics. The majority of short-chain DCA (scDCA) production is based on chemical conversion, synthesised by oxidation under harsh conditions using a metal ion catalyst. Alternatively, oxidative cleavage of cyclic oligomers of butadiene can lead to longer-chain DCAs. However, as the chain length increases, production costs increase significantly due to the generation of numerous by-products that require expensive purification. In addition, the selective ω-oxidation of unsaturated fatty acids by chemical conversion is challenging. Therefore, the synthesis of lcDCAs by microbial production is considered to be more environmentally friendly. LcDCAs are produced by functionalisation of fatty acids or even alkanes via ω-oxidation (see Fig. 3). Inhibition of β-oxidation by knocking out specific genes is necessary to prevent further degradation of lcDCAs to short-chain building blocks. However, by deleting these genes, the use of alkanes or fatty acids as a carbon or energy source is excluded.

In the Bioprocess Development Group, we also pursued metabolic pathway engineering of yeasts, in particular *Canida guilliermondii*, to increase lcDCA production [2]. Although we observed initial improvements in final lcDCA concentrations, further process development was focused on a genetically modified *Candida viswanathii* strain. With the strain—deficient in β-oxidation genes—we achieved a biomass concentration of about 100 g/L of 1,18-octadecene-dicarboxylic acid fed with free oleic acid [3].

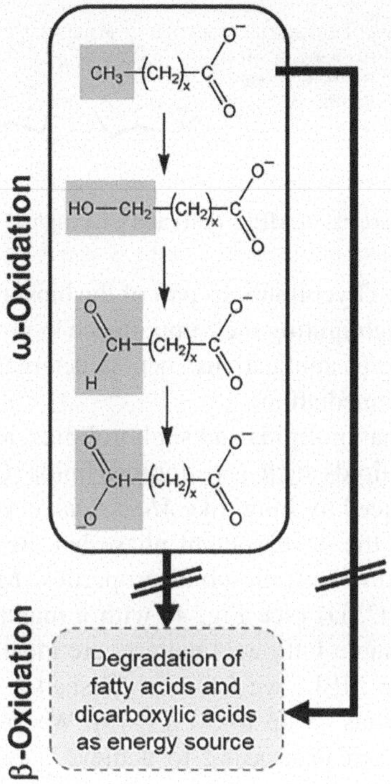

Fig. 3 Metabolic steps in the ω-oxidation. Degradation via β-oxidation is disabled by the knock-out of enzymes

Other current research emphasises the genetic engineering of microorganisms to increase secretion efficiency or to increase lcDCA production in general. Although some processes for the fermentative production of lcDCAs have been commercialised, fermentation remains challenging due to issues such as foaming and substrate and/or product toxicity. We strongly believe that future research should focus on intracellular transport mechanisms (genetic or chemical) of hydrophobic substrates to increase their availability to the microorganism and thus accelerate the ω-oxidation of unconventional substrates. With the increasing interest in integrating system biology approaches such as omics data analysis and metabolic modelling into strain design, new high-performance production strains could be developed to further increase the production of bio-based lcDCAs [4].

4 Biosurfactants from Carbohydrates and Lipids

Chemoenzymatic epoxidation and the production of lcDCAs provide approaches to directly incorporate new functionality into lipids and fatty acids, thereby activating them for further use. In contrast, the microbial production of glycolipids is based on linking parts of the alkyl chain to a hydrophilic sugar core, resulting in a

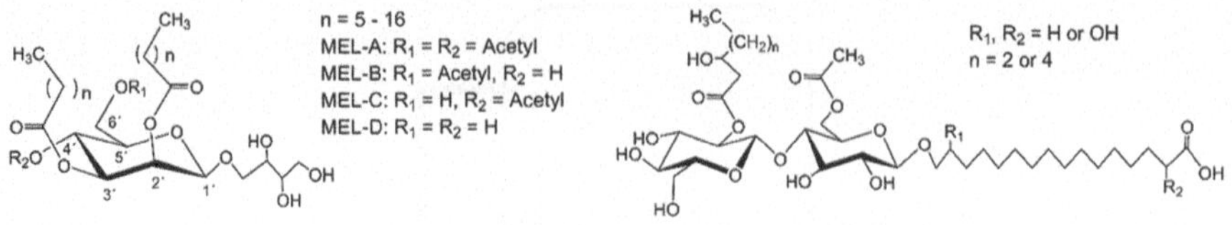

Fig. 4 Basic structures and variants of MELs (left) and CLs (right)

surface-active compound. Glycolipids, as part of the biosurfactant class, show great diversity and potential, highlighting their importance in promoting sustainable production and a wide range of applications such as detergents, cleaners, cosmetics, pharmaceuticals and bioremediation.

Glycolipids such as rhamnolipids and sophorolipids, are well-studied surface-active substances. Glycolipids such as cellobiose lipids (CL) and mannosylerythritol lipids (MEL), produced by fungi like *Moesziomyces* and *Ustilago* sp., are in contrast to these still in the development phase but are promising due to their detergent, emulsifier and/or antimicrobial properties. MELs, produced in four specific variants (A, B, C, D) (see Fig. 4) with a mannosylerythritol unit as a hydrophilic part and variable fatty acid pattern, are ideal for household and personal care products. For MEL, we have established a fed-batch fermentation process with concentrations of up to 50 g/L [5]. We are currently working on optimising the downstream processing to achieve a high-purity biosurfactant (>95% active substance) in a more efficient and resource-saving way. CLs, produced by Ustilaginomycetes, belong to the class of so-called bola-surfactants. They have a cellobiose backbone as a hydrophilic part and two terminal hydrophobic moieties. The CLs exhibit antibacterial and antifungal in addition to the surfactant-specific properties. They are thus suitable for application in crop protection and cosmetics. Here, we have also established fermentation processes with CL concentrations of up to 20 g/L and developed a foam fractionation unit for integrated product recovery [6].

To date, there has been little research into CL production. In contrast to MELs, in addition to fermentation optimisation and the use of complex substrates for fermentation, recent research activities have mainly included application-specific investigations, which have revealed superior properties in various areas. This increases the likelihood of market entry. However, scale-up challenges remain, in addition to improving fermentation yields and reproducibility of product composition. Further research is aimed at improving fermentation productivity and titer, as well as the utilisation of waste streams.

5 L-Malic Acid as Precursor for Polymalate

The microbial production of biosurfactants will soon join the successful industrialisation of fermentative processes for the production of platform and fine chemicals. Success stories range from ethanol to glutamate and citric acid. Although it was also mentioned as a top value-added chemical from biomass, the fermentative production of L-malic acid as a platform chemical is still in its infancy. Our aim has been and continues to be to exploit the potential of L-malic acid as a polymer together with our partners since it has known biomedical effects. Most of the DL-malic acid produced is used in the food and beverage industry as an acidulant and flavour enhancer. DL-malic acid is currently mainly produced from fossil n-butane by hydrogenation of fumaric acid, whereas L-malic acid could be obtained from fossil fumaric acid through fumarase activity. It should be noted that the melting temperature of the enantiomerically pure L form is about 30 °C lower than that of the DL form, which reduces the energy input during melt processing. Microbial production also leads to L-malic acid. It was first patented in 1962, with *Aspergillus flavus* and *A. oryzae* identified as promising strains. The fermentation product (using wild-type strains) could also be classified as 'of biological origin', giving it a higher commercial value. The fermentation process offers advantages such as flexibility in utilising different carbon sources, as demonstrated for fractions derived from flash pyrolysis or organosolv processing of plant materials [7].

At the Fraunhofer IGB, together with our partners, we are contributing to a german project (MALUM) that aims to create a value chain for the production of L-malic acid-based polymers from renewable raw materials and residues using the wild-type strain of *A. oryzae*. It includes the design of a suitable fermentation strategy to avoid using $CaCO_3$ as a buffering agent and the development of a suitable purification process. Scale-up should be carried out using submerged conidia rather than a fermentation cascade. Polymerisation experiments with and without copolymers resulted in the first polymalate samples (see Fig. 5).

Current work on the production of L-malic acid focuses on genetic engineering, particularly of the *Aspergillus* species. In contrast, there is a lack of work on the purification of L-malic acid, although downstream processing represents a significant proportion of future production costs. Resarch concerning the production of L-malic acid focuses on genetic engineering, particularly of the *Aspergillus* species.

Fig. 5 Different Polymalate samples derived from L-malic acid with and without copolymers (kindly provided by HPX Polymers)

6 Polyhydroxyalkanoates with Fine-Tuned Properties

L-malic acid must be polymerised in a subsequent step to produce polymalate. However, some bacteria are able to store polymalate intracellularly as a polymer, with the disadvantage that the composition and polymerisation cannot be optimally controlled. On the other hand, it has the advantage of eliminating a further polymerisation step. In the case of microbially produced polyhydroxyalkanoates (PHAs), the polymer composition can also be very well controlled by the addition of substrates. Discovered by Lemoigne in 1926, PHAs are biopolyesters produced by bacteria as intracellular granules for carbon and energy storage under conditions of nutrient starvation with excess carbon. Due to their biodegradability, biocompatibility and flexibility in physicochemical properties, they have current and future applications in various fields, including medicine and agriculture.

PHAs are composed of (R)-β-hydroxy fatty acids. The most abundant and well characterised are the short-chain length (scl-) PHAs, in particular poly-3-hydroxybutyrate (P3HB), produced by bacteria such as *Cupriavidus necator* or *Azohydromonas lata*. They have high melting and glass temperatures, making them brittle and difficult to process and apply. PHA copolymers, such as poly-3-hydroxybutyrate-co-poly-3-hydroxyvalerate (P3HB3HV), contain two monomers resulting from the feeding of, e.g., levulinic acid, during the fermentation process. These copolymers improve the properties of the material by lowering the melting T_m or the glass transition temperature T_g. Medium-chain-length (mcl-) PHAs are mainly produced by *Pseudomonas* sp. and contain monomers with longer-chain lengths. Our literature review showed that the mcl-PHAs generally have a more rubber-like behaviour due to the low crystallinity. They also have significantly lower melting and glass transition temperatures [8]. Common values for the thermal properties of each polymer are given in Table 1.

The research activity on PHAs is considerable and will be of interest for years to come, as the move towards biocompatible and biodegradable polymers is also inevitable due to social pressure. As there are many ongoing research projects and studies, several topics are being investigated: Once again, genetic modification of PHA-producing microorganisms is a 'hot topic', as is the search for cheap microbial substrates since expensive feedstocks represent a major driver for product costs. Additionally, the variability in monomer composition results in different physicochemical properties and thus also in a variety of studies focusing on application-specific investigations. The high level of interest in PHAs and the large number of studies show the potential, but also that there is still much work to be done: The remaining challenges in PHA production include high costs of downstream processing and low productivity of the microbial strains applied. In addition, the variability of PHA molecular weight and properties, which is on one hand a strength, also hampers commercialisation on the other hand due to a lack of reproducibility.

Table 1 Overview of different polyhydroxyalkanoate basic composition and their resulting thermal properties; P3HO3HN: poly-3-hydroxyoctanoate-co-poly-3-hydroxynonanoate

Name	Structure	Picture	T_m [°C]	T_g [°C]
P3HB	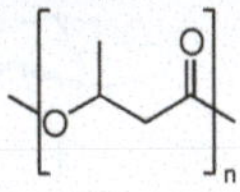		180–190	0–5
P3HB3HV	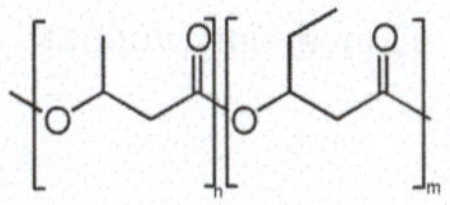	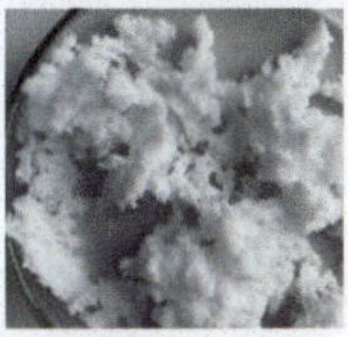	80–90	−13
P_3HO_3HN	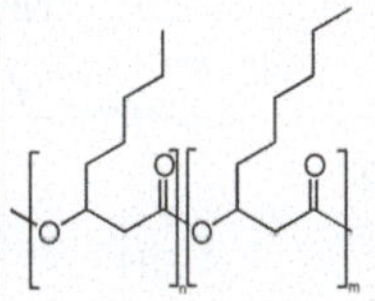		45–60	−30

7 The Biorefinery of Chitin-Rich Biomass

Building polymers through microbial fermentation is one way of harnessing nature's versatility and efficiency. Another alternative is to use 'ready-made' polymers from agricultural or food processing waste streams. This means that the structure and composition of the polymers cannot be manipulated by adaption of processing conditions. Examples include the isolation of starch, cellulose and lignin from various waste streams, as well as chitin.

Chitin, a nitrogen-containing polysaccharide, has been isolated from fishery waste as well as from waste streams from industrial fermentation or insect protein production. It is characterised by its hard, inelastic nature and high hydrophobicity. Despite its abundance, the direct application of chitin is limited due to its insolubility in aqueous phases, its crystallinity and its low reactivity. Chitin already has enormous potential as a substitute for fossil-based polysaccharides, but its conversion to the water-soluble derivative chitosan, which has a higher proportion of free amine groups (see Fig. 6), broadens its utility in many areas: Chitosan's biodegradability, biocompatibility, adsorption capacity and antimicrobial activity make it valuable in agriculture, biomedicine, biotechnology and wastewater treatment.

One of the prerequisites for identifying a suitable chitin source is to have a suitable tool to measure the content. In one of our first papers, we developed a robust

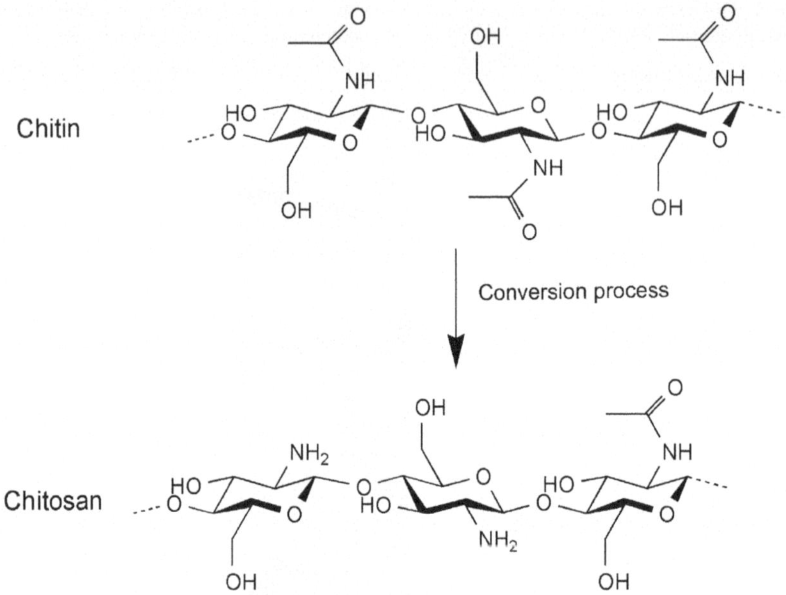

Fig. 6 Chitin could be converted to chitosan which reveals a much higher amount of primary amine groups

analytical tool that allows the researcher to accurately measure the chitin content in insect-based samples [9]. Further work has looked at chitin and chitosan production from different by-products of insect protein production, such as prepupal and pupal exuviae [10–12]. In both cases, we found chitin contents >20% and, after purification and conversion, confirmed that they are a suitable local source of chitin and chitosan. We also focused on the production of chitin and chitosan from the waste streams of other processes, such as fishing [13] or fermentation [14]. The total chemical composition of the residual biomass from fermentation was determined, showing that a chitin purification strategy is required that is very different from that used for fish or insects. However, a chitin and chitosan production process was successfully developed for both, showing that the resulting chitosan is comparable in terms of the amount of free amine groups present, but exhibits significant differences in molecular weight.

With the exception of insect-based sources, current research activity of the scientific community focuses more on the modification of chitin and chitosan and subsequent applications rather than the establishment of new isolation processes from other sources. Although a large number of studies have been published on process development for the production of chitin and chitosan from various sources, commercial implementation is in comparison poor. This may be due to the chitin-to-chitosan conversion process, which is energy-and chemically intensive and may need to be refined with regard to the implementation of biotechnological purification tools.

8 Conclusion

At the International Workshop on Circular Bioeconomy, the Bioprocess Development Group of the Fraunhofer IGB presented its latest findings and experiments on various facets of industrial biotechnology and biorefinery. The group has carried out innovative research on the use of plant oils for the production of epoxides and biosurfactants, the production of microbial monomers (such as malic acid) and polymers (such as polyhydroxyalkanoates), and the direct extraction of chitin and chitosan from agricultural waste streams.

In conclusion, the Fraunhofer IGB team's contribution to the Circular Bioeconomy Workshop has shed valuable light on efforts to move away from petroleum-based processes towards a sustainable bio-based industry. The research presented at the workshop reinforces the role of industrial biotechnology as a key driver for a sustainable bioeconomy. Although several challenges remain, the broad outlook towards innovative solutions, cross-cutting technology applications and continuous exploration of bio-based alternatives lays the groundwork for future breakthroughs.

Acknowledgements Financial support for this chapter has been provided by the Bioeconomy International program 2020, supported by the German Ministry of Research BMBF under funding no. 031B1239, project CiBELESα. We also like to thank the German Ministry of Food and Agriculture BMEL for funding of the L-malic acid and polymalate research within the project Malum under funding no. 2220NR272.

The authors would like to thank Alexander Beck and Pravesh Tamang for reviewing several chapters and Birk Achenbach, Lisa Kern and Steven Leonhardt for the provision of data and pictures.

References

1. Haitz F et al (2018) Applied biochemistry and biotechnology. Springer, Cham, pp 13–33
2. Werner N et al (2017) Biotechnology letters, pp 428–439
3. Zibek S et al (2009) Chemie Ingenieur Technik, pp 1797–1808
4. Werner N, Zibek S (2017) World J Microbiol Biotechnol:194
5. Beck A et al (2022) Front Bioeng Biotechnol:913362
6. Oraby AH et al (2023) Discover chemical engineering
7. Dorsam S et al (2017) Biotechnology for biofuels
8. Hahn T et al Engineering in Life Sciences n/a: 2300211
9. Hahn T et al (2018) J Sci Food Agric 98:5069–5073
10. Guarnieri A et al (2022) Sci Rep 12:8084
11. Hahn T et al (2020) J Biotechnol 310:62–67
12. Hahn T et al (2022) Waste and Biomass Valorization 13:1993–2008
13. Hahn T et al (2024) J Appl Polym Sci:e54789
14. Krake S et al (2024) Biotechnol Bioprocess Eng 29:942–954

Biorefineries Based on Coupling of Wastewater and Microalgae: Developments of Algal Biotechnology

Ulrike Schmid-Staiger and Konstantin Frick

Highlights

- Microalgae production is a link between the sector of wastewater treatment, recycling of nutrients, waste gas CO_2 and biorefineries as part of the circular bioeconomy.
- The state of the art of algae biotechnology is shown as a cross-sectoral technology to produce substances from waste streams, such as CO_2 and nutrient-rich digestate, for application in different sectors.
- Key principles for sustainable and economic algae biotechnology are shown with regard to photobioreactor (PBR) technology and the necessary energy and cost reduction.

1 Introduction

Where will the raw materials of the future come from? How do we reduce waste and combat climate change at the same time? These questions are at the centre of the circular economy. One of the essential goals of the circular bioeconomy is to minimize waste through the total use of biomass, which is achieved via the biorefinery approach. This approach ensures that in the case of microalgae biomass, all components are utilized, thereby maximizing the resource-efficient use of biomass and enhancing the overall economic viability of microalgae technology.

U. Schmid-Staiger (✉) · K. Frick
Fraunhofer Institute for Interfacial Engineering and Biotechnology IGB, Innovation Field Industrial Biotechnology, Microalgae Biotechnology—Development Group, Stuttgart, Germany
e-mail: ulrike.schmid-staiger@igb.fraunhofer.de

A. Meneses-Jácome et al. (eds.), *The Circular Bioeconomy in Industry*, Waste as a Resource, https://doi.org/10.1007/978-3-031-84359-4_2

The circular bioeconomy aims to promote the use of renewable biological resources across various sectors, including biopolymer production, food and feed production and bioenergy. By harnessing the capabilities of microalgae, we can produce valuable biomass for human-related applications such as pharmaceuticals, cosmetics, food and feed. Microalgae technology has the potential to make significant contributions to the circular economy and the transformation to a bioeconomy and thereby mitigate climate change. One of the key benefits of microalgae is their high capacity for CO_2 capture by use of CO_2 in photosynthetic growth and biomass production. They can fix CO_2 from various sources, including the atmosphere, industrial exhaust gases or CO_2 from biomethane production. Microalgae have higher growth rates and CO_2 fixation capabilities compared to conventional plants, making them more effective for carbon sequestration. Microalgae can be integrated into wastewater treatment processes or downstream processes of wastewater treatment such as the fermentation of sewage sludge in biogas plants. They are capable of nutrient recovery from wastewater or derived sources, such as digestate from biogas plants, converting these nutrients into valuable biomass that can be converted into biofuels, biostimulants, animal feed and other bio-based products. This ability to recycle nutrients enhances circularity and resource use and at the same time reduces the burden on the environment by assimilating their excess.

Sustainable algae production is characterized by several key factors that collectively enhance its environmental and economic viability. One of the primary characteristics is high biomass productivity, achieved per unit area and volume, which translates into low land usage. This efficiency is particularly advantageous as algae can be cultivated in photobioreactors, optimizing space and resource utilization. The efficient use of resources is another advantage of microalgae. They require minimal land and have a low water footprint compared to traditional crops [1]. However, it is important to note that microalgae cultivation can require high energy input, but this environmental impact can be reduced if renewable energy sources are used. To ensure sustainable algae production, it is essential to focus on the production of algae biomass and specific ingredients in resource- and cost-optimized closed scalable systems. These systems should be designed to be energy efficient, reducing the overall environmental impact. Moreover, adopting a biorefinery approach allows for the total utilization of biomass, producing a wide range of applications from biofuels and bioplastics to pharmaceuticals and nutraceuticals. This holistic approach ensures that all components of the algae are used, minimizing waste and maximizing economic returns.

In conclusion, sustainable algae production is characterized by high biomass productivity, efficient resource use, low greenhouse gas emissions, and the recyclability of nutrients and water. By adopting energy-efficient processes and scalable closed systems and leveraging the biorefinery approach, algae production can be made both sustainable and economically viable, contributing to a more resilient and eco-friendly future. This contribution presents the impact of key principles needed for sustainable algae technology as part of the circular bioeconomy. The opportunities offered by algae biotechnology in a biorefinery concept based on wastewater will be outlined.

2 Microalgae and Circular Bioeconomy

Circular bioeconomy aims to promote the use of renewable biological resources across various sectors, including biopolymer production, food and feed production up to bioenergy. By harnessing the capabilities of microalgae, microalgae technology has a high potential to make significant contributions to circular bioeconomy by closing nutrient cycles and integrating side streams in algae production for value-added products. One of the essential goals of the circular bioeconomy is to minimize waste through the total use of biomass, which is achieved via a biorefinery approach. This approach ensures that all components of the microalgae are utilized, thereby maximizing the resource-efficient use of biomass and enhancing the overall economic viability [2–4].

One of the key benefits of microalgae is their high CO_2 capture capacity, especially by coupling with other processes. Microalgae are suitable for the 'utilization' of CO_2 from industrial point sources. These can be CO_2 streams not only from biogas production but also from fermentation processes in food production or process-related CO_2 emissions, such as those from the cement industry, making them effective in mitigating climate change (Fig. 1). CO_2 fixation by microalgae in photobioreactors presents a promising solution for reducing greenhouse gas emissions efficiently. About 1.8 kg of CO_2 is converted to 1 kg of microalgae biomass. CO_2 conversion is directly coupled to space-time yield or biomass productivity per area or per volume. Therefore, CO_2 fixation rate and efficiency depend on the production system used, i.e. open systems or closed photobioreactors, outdoor with sunlight or indoor with artificial illumination, as these are factors that strongly influence microalgae growth and biomass productivity.

Nutrient recycling and wastewater treatment are other areas where microalgae can contribute to circular bioeconomy. A wastewater treatment plant can be seen as a biorefinery in an urban or industrial environment. The question always arises as to where resources can be recovered. Wastewater treatment plants play a major role in this context, as many material flows come together. It is a process engineering

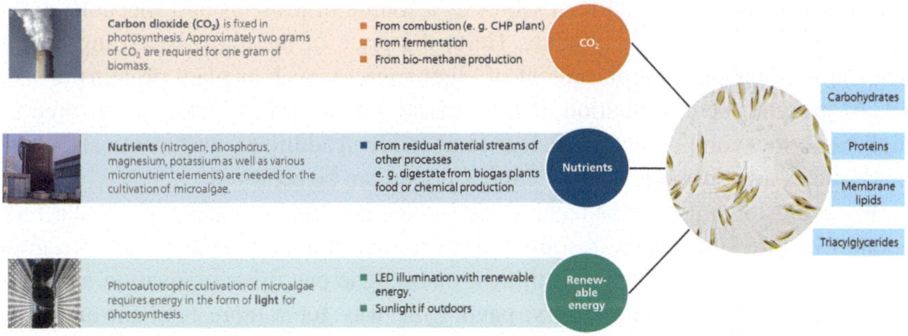

Fig. 1 Integration of side streams from different origins into algae cultivation to close nutrient cycles for a circular bioeconomy

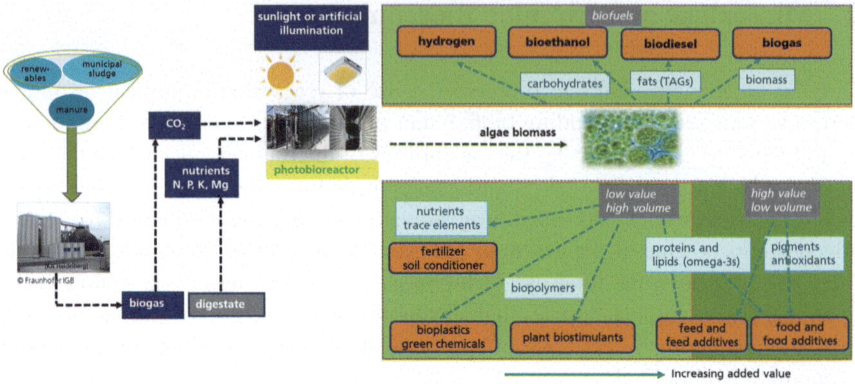

Fig. 2 Coupling of anaerobic digestion as a resource for CO_2 and nutrients with microalgae biomass production in a circular bioeconomy set-up. Possible microalgae ingredients for application in various sectors such as food, feed, materials, energy and agriculture. The substrates used in the AD determine the possible use of the ingredients in the various sectors, in accordance with the relevant regional regulations

principle that the more concentrated the materials are, the better they can be recovered. The part of the wastewater treatment plant where the nutrient concentrations are highest is the sludge digestion. After anaerobic digestion, main nutrients such as nitrogen are available as ammonium, the best source of nitrogen for many microalgae species [5, 6]. Coupling anaerobic digestion as a resource for CO_2 and nutrients with microalgae biomass production in a circular bioeconomy set-up has a high potential. However, the substrates used in the anaerobic digestion determine the possible use of the ingredients in the various sectors, in accordance with the relevant regional regulations, such as those for food and feed.

Moreover, microalgae can be a source of bio-based materials for various sectors (Fig. 2), including biofuels, bioplastics and other biomaterials—the so-called high-volume, low-value products. Microalgae can contribute to soil improvement by adding organic matter, enhancing soil structure and providing nutrients, thereby promoting sustainable agriculture practices. This also includes microalgae that could be used in the animal feed sector as a source of proteins and speciality fatty acids, such as omega-3 fatty acids. Here, the price of the algae-based ingredient determines their use or replacement of other sources, such as plant proteins or fish oil. At present, the contribution of microalgae is focussed primarily in the area of high value-low volume sectors such as food and feed additives, nutraceuticals, pharmaceuticals and cosmetics, where higher prices can be achieved.

In conclusion, microalgae technology has the potential to make the circular bioeconomy more sustainable, resource-efficient and climate-friendly. By leveraging their unique properties and capabilities, microalgae can drive innovation and sustainability across multiple sectors, paving the way for a more resilient and eco-friendly future.

3 Photobioreactors

To utilize the potential of microalgae in a circular bioeconomy, sustainable processes must be developed, thereby using photobioreactors or open systems. To achieve this, there are requirements for the reactor system that makes sustainable algae production possible in the first place. The cultivation systems for algae—such as open ponds, bubble columns, tubular systems and flat plate airlift reactors—differ significantly in terms of volume, area, light energy and process control. In summary, open ponds are cost-effective and scalable but offer limited process control and higher contamination risks, resulting in a low biomass concentration, low productivity and low quality of the produced biomass. Bubble columns, tubular systems and flat plate airlift reactors provide better process control, efficient CO_2 utilization and lower contamination risks, making them suitable for more controlled and higher productivity operations. However, they involve higher initial costs and more complex operational requirements.

The challenge in algae cultivation is process intensification, a balancing act between sustainability and profitability. One of the key aspects of this approach is optimized light utilization. Photobioreactors allow for precise control over the light regime, which significantly enhances productivity and light conversion efficiency. By providing algae with the optimal amount of light, these systems maximize photosynthetic activity and biomass production. Water and nutrient recycling is another critical component of process intensification. Photobioreactors enable the efficient use of water and nutrients, with low evaporation losses and no nutrient loss through leaching. This closed-loop system ensures that resources are used more effectively, reducing both costs and environmental impact. Efficient CO_2 capture is integral to the sustainability of algae production.

In photobioreactors, CO_2 utilization is closely linked to the light regime and nutrient supply, ensuring that the CO_2 introduced is completely converted into biomass. This not only enhances the growth rate of the algae but also allows for the use of CO_2 emissions from industrial processes, turning a waste product into a valuable resource. Moreover, the high level of control and automation possible in photobioreactors leads to more stable and efficient processes. Automation of the lighting regime and feeding management, particularly in artificially lit PBRs, results in more energy-efficient processes. To achieve this, the use of sensors and their integration into the control system is a prerequisite for automated processes [7]. The ability to tailor production parameters to specific needs results in optimized output and quality.

Year-round production is another significant advantage of photobioreactors. Unlike traditional farming, which is often seasonal, photobioreactors can operate continuously, providing a consistent supply of biomass and specific ingredients. Regional climate and energy costs determine the light source to be used. Space efficiency and scalability are additional benefits of photobioreactors. Their modular and scalable nature allows for the efficient use of space, making them suitable for diverse environments, including urban and industrial settings. This flexibility ensures that algae cultivation can be integrated into different types of environments, further enhancing its sustainability and economic viability.

In conclusion, process intensification through the use of photobioreactors offers a balanced approach to achieve both sustainability and profitability in algae production. By optimizing light utilization, recycling water and nutrients, capturing CO_2 efficiently, leveraging advanced control and automation, ensuring year-round production and maximizing space efficiency and scalability, this method ensures that algae cultivation is both economically viable and environmentally friendly.

4 Cutting Costs Is Mandatory for Algae Production

Large-scale application of microalgae cultivation is currently limited by the costs for production, both capital expenditures (CAPEX) and operating expenditures (OPEX). In LED-illuminated PBRs, 75% of the cultivation costs are lighting and 15% is CO_2 supply; cost reduction for these main factors is therefore mandatory for microalgae biomass production [8]. Dissolved CO_2 levels have to be maintained high for high biomass productivity, and at the same time, CO_2 losses via the off-gases have to be reduced. This can be achieved on the one hand by recirculating the off-gas and by increasing the mass transfer of CO_2 in the culture by the reactor design and intermixing. The recirculation of the off-gas has to be controlled by O_2 sensors so that CO_2 fixation is not impaired by excessively high dissolved O_2 concentrations in the medium [9]. Elevated O_2 concentrations and the corresponding increase in the ratio of O_2:CO_2 common in photobioreactors led to a reduction of growth and biomass yield on light of 20–30%. Other factors which influence the solubility and mass transfer of CO_2 in the culture medium are the pH of the medium, which can affect the chemical equilibrium and availability of CO_2 for fixation, and temperature, which affects both the solubility of CO_2 and the metabolic rates of the microalgae. The size and residence time of gas bubbles in the culture volume are additional factors that determine the efficiency of CO_2 mass transfer. Smaller bubbles with longer residence times enhance the contact between CO_2 and the microalgae, thereby improving the overall fixation rate. These factors must be taken into account during the process and reactor development.

The main advantage of the use of artificial light is the better regulation of the photosynthetic photon flux density (PFD) and light spectra, which may result in better biomass quality and productivity [10–13]. The higher energy costs are partially compensated by a much higher biomass productivity compared to cultivation with sunlight, as well as a high light yield. Both depend on the biomass concentration in the reactor volume. The high energy costs involved in production with LED illumination can be reduced in several ways:

- The use of efficient LEDs that have now been developed for horticulture. The PFD per Watt that can be generated has been increased to up to 4 μmol photons per Watt of electricity in recent years.

- Designing the process in such a way that a high light yield or photosynthetic efficiency is achieved (PE in g DW/mol photons). The light distribution in the reactor volume is the decisive factor here, which is particularly dependent on the biomass concentration. This can be achieved by reducing the light path by irradiating from both sides in the case of a flat-plate reactor or by efficient intermixing [14, 15].
- Adaptation of PFD to increasing biomass concentrations to keep the light per g biomass constant results in a high PE and a high volumetric productivity [16]. This can be achieved by the integration of an optical density (OD) sensor into the control system.

Moreover, the high level of control and automation possible in photobioreactors leads to more stable and efficient processes. The ability to tailor production parameters to specific needs results in optimized output and quality. Automation also reduces labour costs and minimizes human error, further enhancing profitability. Processes geared toward energy efficiency offer another avenue for cost reduction. Using 'green electricity' for the production's own electricity needs can substantially lower electricity costs. In conclusion, cutting costs in algae production necessitates a multifaceted approach. By optimizing process control, enhancing energy efficiency, adopting a biorefinery approach and implementing efficient processing methods, algae production can become more economically viable. These strategies not only reduce costs but also increase the added value of the final products, making algae cultivation a sustainable and profitable venture.

5 Application of Energy-Efficient Algae Technology

Wastewater treatment plants are an ideal example of how the ideas of the circular economy and bioeconomy can be realized by producing new biomass from waste such as CO_2 together with excess nutrients, which helps to reduce the burden on the environment and the climate [17]. The above-mentioned principles for energy-efficient processes and reactor development were demonstrated in a biorefinery project using nutrients and CO_2 from the sludge digestion process (Fig. 3). Nutrient-rich liquid digestate with NH_4 concentrations up to 600 mg/L, separated insoluble struvite (magnesium ammonium phosphate) from a digestate refinery approach and CO_2 from biomethane production were used to produce ß-glucan-rich algal biomass as a plant biostimulant or soil conditioner. For this realization, a 5 × 25L photobioreactor module for algae cultivation with integrated off-gas recirculation to increase CO_2 utilization was built (Fig. 4). The flat plate photobioreactors were illuminated on both sides by LEDs, and by the integration of sensors into the control system, the light intensity was automatically adapted to the increasing biomass concentration in the reactor volume. Nutrient-rich digestate was provided in the culture medium and the required N:P ratio was adjusted with the addition of struvite. This set-up resulted in a high biomass productivity (average of 1.5 g DW $l^{-1}d^{-1}$ with a growth phase

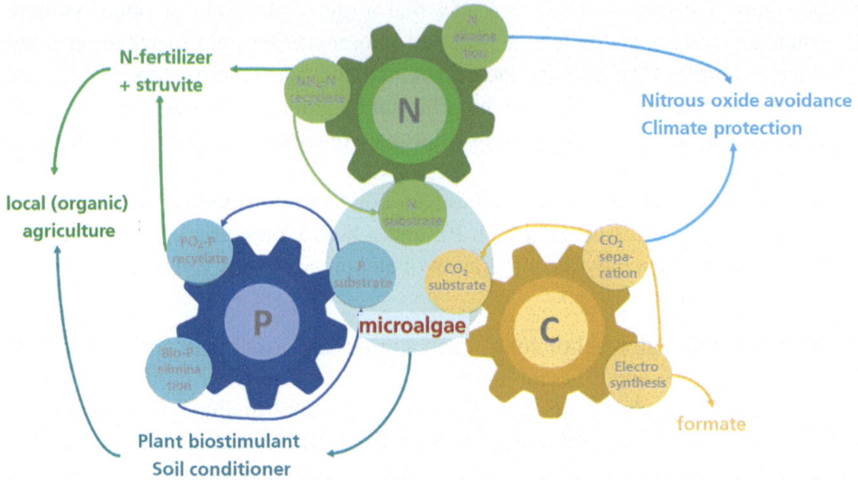

Fig. 3 Microalgae production as part of a wastewater treatment plant-based biorefinery. Microalgae consume CO_2, ammonium and phosphate; the biomass can be used as plant biostimulants or soil conditioners, thereby avoiding nitrous oxide production and CO_2 emissions at the wastewater treatment plant

Fig. 4 Photobioreactor set-up in a wastewater treatment plant environment using CO_2 from biomethane production and nutrient-rich liquid digestate from anaerobic sludge digestion

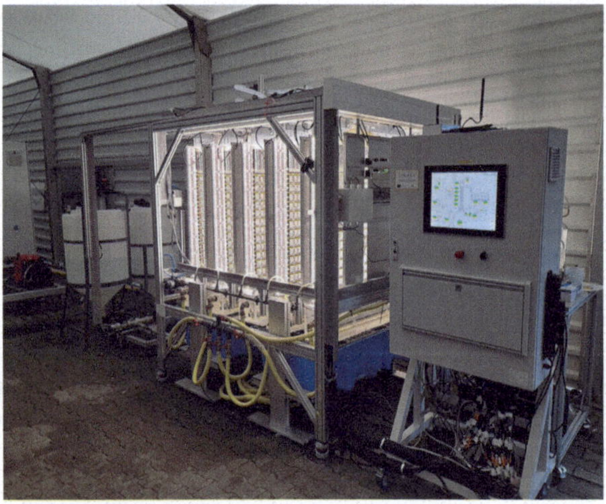

between 2 and 6 g DW/L) and high light yield (0.6–1 g DW/mol photons), thereby fixing 90% of the added CO_2.

In conclusion, microalgae in photobioreactors offer a highly efficient method in anaerobic sludge digestion for CO_2 mitigation. Nutrients in digestate were completely consumed and support good growth in photobioreactors. This technology cannot only be used on wastewater treatment plants but also for other industrial CO_2-containing off-gases and nutrient-rich sidestreams, making this technology a potent tool for addressing climate change.

6 Conclusion

Microalgae technology has the potential to make significant contributions to the circular economy and the transformation to a bioeconomy, thereby mitigating climate change. The realization of the idea of a sustainable circular bioeconomy with microalgae stands and falls with the production process, which must be energy-efficient, robust and, above all, scalable in energy-efficient photobioreactors. The necessary process intensification can be achieved by the following:

- Improving light utilization by offering precise control of the light regime and the biomass concentration. This is possible with artificial illumination and a process control system with integrated sensors.
- Efficient CO_2 capture by microalgae and complete conversion into valuable biomass.
- Water and nutrient recycling via the use of liquid digestates from anaerobic digestion of residues, such as municipal sludge.
- Scalable photobioreactors suitable for environments such as urban and industrial settings.

In conclusion, process intensification with photobioreactors offers a balanced approach to achieve both sustainability and profitability in algae production.

Acknowledgements Financial support for this chapter has been provided by the Bioeconomy International program, a program supported by the German Ministry of Research BMBF under funding no. 031B1239, project CiBELESα.

References

1. Moomaw W, Berzin I, Tzachor A (2017) Ind Biotechnol 13(5):234–243. https://doi.org/10.1089/ind.2017.29102.wmo
2. Ruiz J et al (2016) Energy Environ Sci 9:3036. https://doi.org/10.1039/c6ee01493c
3. Samoraj M et al (2024) Biocatal Agric Biotechnol 58(2024):103224. https://doi.org/10.1016/j.bcab.2024.103224
4. Acién Fernández FG et al (2021) New Biotechnol 61:99–107. https://doi.org/10.1016/j.nbt.2020.11.011
5. Meiser S-SU, Trösch W (2004) J Appl Phycol 16:215–225. https://doi.org/10.1023/B:JAPH.0000048507.95878.b5
6. Schmid-Staiger U et al (2009) Chemie Ingenieur Technik 81(11):1783–1789
7. Yeh Y-C et al (2023) Front Bioeng Biotechnol 11:1116735. https://doi.org/10.3389/fbioe.2023.1116735
8. Stange M et al (2023) Target costing as an approach to reduce costs in closed-loop agriculture systems - application for the cultivation of Algae in photobioreactors. Procedia CIRP 116(2023):372–377. https://doi.org/10.1016/j.procir.2023.02.063
9. Kliphuis AMJ et al (2011) Biotechnol Bioeng 108(10):2390–2402. https://doi.org/10.1002/bit.23194
10. Kubín Š et al (1983) Biochem Physiol Pflanz 178(2–3):193–205. https://doi.org/10.1016/S0015-3796(83)80032-8

11. Blanken W et al (2013) Algal Res 2:333–340. https://doi.org/10.1016/j.algal.2013.09.004
12. Amaro HM et al (2020) J Appl Phycol 32:3605–3618. https://doi.org/10.1007/s10811-020-02212-2
13. Karagülle G, Telli M (2024) Biotechnol Prog 40:e3462. https://doi.org/10.1002/btpr.3462
14. Cuaresma M et al (2009) Biotechnol Bioeng 104:2. https://doi.org/10.1002/bit.22394
15. Zijffers JWF et al (2010) Mar Biotechnol 12:708–718. https://doi.org/10.1007/s10126-010-9258-2
16. Frick K et al (2023) Microbiology Open 12:e1378. https://doi.org/10.1002/mbo3.1378
17. Khan S et al (2023) Curr Opin Green Sustain Chem Bd 41, S:100797. https://doi.org/10.1016/j.cogsc.2023.100797

Progress on "Piñafibre" Project: Unlocking the Potential of Dual-Purpose Crop by the Utilization of Pineapple Leaf Fibers for Bio-Based Textiles

Isabel Moreno, Mauricio Fernando Martínez, Daniel Ortiz, Yohana Mora, Oscar Paredes, Yaneth Patricia Ramos, Carol Puentes, Kevin Rodríguez, Marie Bolger, and Tobias Wojciechowski

Highlights

Pineapple leaf fiber extraction (PALF) enables a pathway for crop residue management with a circular bioeconomy approach.

1 Introduction

Pineapple, or Ananás (*Ananas comosus* (L.) Merrill), is the most important american fruit and the third tropical fruit after banana and mango [1]. It is cultivated in all tropical and subtropical regions of the world. According to FAO, its global production in 2022 was 29.4 million tons. In Colombia, this productive sector has increased in recent years, reaching 22,870 hectares cultivated in 2022 and a production of 919,709 tons [2]. To safeguard the great genetic diversity of this species, germplasm collections have been established in various regions worldwide, as their characterization constitutes an ideal tool for studying their variability and genetic diversity, as well as providing input for genetic improvement programs [3].

In Colombia, there is a germplasm bank of the genus *Ananas* spp., which is part of the National Germplasm Banks System for Food and Agriculture (SBGNAA), established at Agrosavia, Palmira Research Center. Pineapple is mainly cultivated

I. Moreno (✉) · M. F. Martínez · D. Ortiz · Y. Mora · O. Paredes · Y. P. Ramos · C. Puentes · K. Rodríguez
Corporación Colombiana de Investigación Agropecuaria, Agrosavia, Centro de investigación Palmira, Palmira, Valle del Cauca, Colombia
e-mail: mimoreno@agrosavia.co

M. Bolger · T. Wojciechowski
Forschungszentrum Jülich GmbH (FZJ), Jülich, Germany

A. Meneses-Jácome et al. (eds.), *The Circular Bioeconomy in Industry*, Waste as a Resource, https://doi.org/10.1007/978-3-031-84359-4_3

for fruit commercialization, which generates a large amount of plant residues or biomass since one plant produces approximately 40–50 leaves or 2.3 kg of plant residue equivalent to 250 metric tons.ha^{-1} of fresh plant residues [4].

Internationally, several research studies have been developed to find alternatives for the management of these agricultural residues with the purpose of reintegrating them into the economic cycle as by-products for final consumption or as inputs for other systems. Regarding the utilization of pineapple harvest residue, it is estimated that between 2% and 7% of the leaf corresponds to fiber that can be used as raw material for various industries due to its characteristics of whiteness, brightness, softness, salt resistance, tensile strength, and capacity to retain water and dyes [5]. In recent years, the use of natural fibers has captured worldwide interest as a potential substitution for synthetic materials due to their physical and mechanical properties and their abundant availability [6]. Internationally, there are already initiatives for the production on a commercial scale of by-products with pineapple crop residues, as presented by Vieira et al. (2022) [7].

In Colombia, some efforts have been made to create companies that use the circular bioeconomy model to utilize vegetable residues from various crops to generate added value for producers, focusing on specialized ecological markets [8]. When considering the market projection of plant fiber, it is important to highlight that each industry requires specific fiber properties; therefore, the physicochemical and mechanical characterization of the fiber is important for evaluating its industrial potential [9]. The objective of this phase of the project is to assess the genetic variability of pineapple germplasm bank accessions through phenotypic characterization and explore their attributes associated with fiber production for agro-industrial use, aiming to contribute to the generation of alternatives for the utilization of pineapple crop biomass.

2 Material and Methods

2.1 Morphological Characterization and Fiber Extraction

The Pineapple Germplasm Bank of the Colombian Corporation for Agricultural Research AGROSAVIA, Palmira Research Center, is located in the municipality of Palmira in the department of Valle del Cauca. For this study, plant material from 75 pineapple accessions conserved ex situ was characterized and collected. For morphological characterization, 17 plant characters were recorded at the reproductive stage including ten quantitative ones such as plant height (PH), stem diameter (SD), number of leaves (NL), leaf thickness (LT), fresh leaf weight (LWf), dry leaf weight (LWd), leaf length (LL), leaf width at the base, middle, and apex (LWoB, LWoM, and LWoA), and seven qualitative ones including plant posture (PPl), peduncle length (PeL), piping (PI), presence of trichomes (TRI), presence of spines on leaf edges (SpL), spine density (SpD), and spine position on the leaves (SpP). The variables were selected based on standardized descriptors for pineapple described by the International Plant Genetic Resources Institute [10] and the International Union for the Protection of New Varieties of Plants [11].

Before defibration, ten leaves per accession were randomly selected, including at least one leaf from each plant replication. Each leaf was measured for weight (LWf), length (LL), width (LW), and leaf thickness (LT). Mechanical fiber extraction was carried out at a farm in the municipality of Totoró during the second semester of 2022, in the Cauca department, Colombia. The plant material used for fiber extraction corresponded to 74 out of 75 pineapple accessions, which had available leaves suitable for extraction. Additionally, the fiber obtained per replication was measured for length (FL) (cm) and weight (FW) (g). The fiber extraction percentage (Fp) (%) was estimated, calculated by dividing the weight of the dry fiber (FW) by the weight of the fresh leaves (LWf), and for fiber production (FP) weight/leaf was calculated. The quantitative morphological characters data were subjected to principal component analysis (PCA) and cluster analysis, while multiple correspondence analysis (MCA) was conducted for the qualitative variables. Additionally, variables recorded during fiber extraction such as FL, FP, and Fp along with morphological leaf variables including LWf, LL, LW, and LT were analyzed using correlation matrices and descriptive statistics. All analyses were performed using the SAS® 9.4 (Statistical Analysis System) software for Windows.

2.2 *Socioeconomic Analysis*

The economic costs of managing 1 hectare under three waste management options: (1) drying and burning, (2) mechanization, and (3) manual collection for fiber extraction were estimated through consensus with pineapple producers located in the Northern Cauca region during a workshop. Income from 1 hectare of fiber sales was estimated and compared with the associated costs for processing residual leaves.

3 Results and Discussion

3.1 *Morphological Characterization and Fiber Extraction*

The evaluation of qualitative morphological characters in *Ananas* accessions showed high variation. These characters are highly heritable, as they are less influenced by the environment. Leaf characters, such as SpL, SpD, and SpP, were highly discriminatory and managed to separate the accessions into four groups with a high degree of heterogeneity with a dissimilarity coefficient of 0.49. Group 1 had the highest composition with 60 accessions and the formation of three clearly differentiated subgroups (SGI, SGII, and SGIII). Group 2 had 12 accessions, group 3 had six accessions, and group 4 consisted of a single accession, which behaves as a contrast group. Variation in qualitative characters has been reported by several authors in the characterization of pineapple germplasm from the central and southern regions of Benin (West Africa) [12] and Indonesia [13]. Similarly, quantitative characters of plants and leaves showed high variability. Those characters are measurable traits generally influenced by the environment and the action of several genes, where the effect of an individual gene is difficult to detect. Of the ten evaluated characters, three had a coefficient of variation (CV) less than 20%, indicating high homogeneity in the variable within the population. LWoA was the most variable, while SD and LWoM were the least variable. Based on principal component analysis, it was evidenced that the quantitative characters evaluated can explain 87.4% of the total variability in four components, where PH, DLL, LWoB, and LWoM are the major contributors.

The evaluated characters also showed a great variation in pineapple diversity in Cuba [14] and Indonesia [13]. In principal component 2 (PC2), the variables with the greatest contribution to the variation are those related to biomass accumulation, such as LWf, LWd, and relative water content (RWC). In principal component 3 (PC3), LWoA is the major contributor. The quantitative morphological characters formed five defined groups in the pineapple accessions, which also corresponds to the structure generated by qualitative characters and confirms their high conserved phenotypic variability. In addition, the descriptive analysis of the quantitative characteristics of defibrated leaf accessions showed that the variables FP, Fp, and LWf presented the highest variability with coefficient of variation values of 35%, 29%, and 39% and average values of 1.5 g, 3.5%, and 39 g, respectively. Regarding the correlation analysis, the highest coefficients were found in FL followed by LWf, and finally FP. FL had a directly proportional relationship with LL, LWf, and FP with correlation coefficients of 0.79, 0.65, and 0.46, respectively. LWf showed a higher relationship with LL, LW, and FL as correlation coefficients of 0.69, 0.65, and 0.65, respectively, were found.

These preliminary results suggest that the mechanical extraction method, in addition to its efficiency in terms of extraction time, is less destructive and preserves fiber length according to the initial length of the processed leaf. Finally, FP was associated with the variables Fp, FL, LL, and LWf, with correlation coefficients of 0.48, 0.46, 0.32, and 0.31, respectively. Based on our preliminary results, the variables with the highest variability were FL, LWf, and FP, which are important attributes for the utilization of crop harvest residues. FP had a directly proportional relationship with the morphological variables of LL and LWf. In turn, LWf had a strong association with variables associated with leaf dimensions in terms of length and width. These results are consistent with those found by Ortiz-González et al. [15], who reported that plant fiber production is influenced by the number and size of leaves. These results did not allow us to determine a clear correlation between the evaluated morphological variables and fiber production. However, it should be considered that different accessions from a germplasm bank are being analyzed, so the analysis should be complemented to evaluate the hypothesis that this variable could be associated with the genetic component of the materials and not with their morphological characters. The diversity of qualitative traits in pineapple genetic materials has been previously reported by several authors [16].

3.2 Socioeconomic Analysis

The results suggested that the option of manual collection and fiber extraction from pineapple leaves yields a positive return on investment, making it economically viable. Taking into account the economic costs associated with harvest residue management, the most expensive alternative is manual leaf collection, explained by the intensive labor required by this option, while the cheapest option from an economic point of view is Desiccate and Burn. However, this preference overlooks the social and environmental costs. Additionally, the assessment of utilizing pineapple leaves for fiber extraction in a small-scale model showed that it is financially viable, based on the defined assumptions regarding fiber production, market prices, and production costs considered in this analysis. However, a complementary social and environmental valuation analysis of this utilization is currently in progress to assess from a multidimensional perspective, the economic feasibility of the pineapple production system.

4 Conclusion

The high percentages of variation in morphological traits found in this study indicate the high genetic diversity of this genus in the National Germplasm Bank. These results could be obtained because the accessions were collected from different geographic regions, demonstrating processes of genetic segregation of their progenitors and low levels of domestication. For this reason, this study also contributes to the baseline understanding of fiber production in *Ananas* genotypes in Colombia, which has been scarce thus far.

Acknowledgments The authors would like to thank the Ministry of Science, Technology, and Innovation of Colombia for funding project 452-2021 through the "FRANCISCO JOSÉ DE CALDAS" NATIONAL FUND FOR SCIENCE, TECHNOLOGY, AND INNOVATION.

References

1. Rodríguez D et al (2007) Rev Fac Agron Luz 24(1):20–26
2. FAO (Food and Agriculture Organization o the United Nations) https://www.fao.org/faostat/en/#home. Consulted in April 2022
3. Núñez-Colín CA, Escobedo-López D (2011) Agronomía Mesoamericana 22(2):415–427. (In Spanish)

4. Chen A et al (2020) Biomass Bioenergy 141:105675
5. Jawaid M et al (2020) Pineapple leaf fibers: processing, properties and applications. Springer, Cham, pp 1–321
6. Jain J, Sinha S (2022) J Nat Fibers 19(15):10031–10052
7. Vieira IMM et al (2022) Bioenergy Res 15(2):924–934
8. Luna G et al (2009) Dyna 76(159):145–151
9. Seiko J, Salim R, Ammayappan L (2016) J Nat Fibers 13(3):362–373
10. IBPGR (1991) Descriptors for pineapple. The International Board for Plant Genetic Resources, pp 11–33
11. UPOV (2013) Guidelines for the conduct of tests for distinctness, uniformity and stability—pineapple (Ananas comosus (L.)) (Merr). UPOV
12. Adje CAO et al (2019) Fruits 74(4):167–179
13. Rosmaina. et al (2021) Biodiversitas 22(9):3854–3862
14. Rodríguez-Alfonso D et al (2020) Hortic Sci 47(1):28–35
15. Ortiz-González D et al (2021) Centro Agrícola 48(1):5–13. (In Spanish)
16. Zizka A (2020) Divers Distrib 26(2):183–195

Use of Basidiomycetesin the Valorization of Agro-Waste as an Alternative in Implementing the Circular Bioeconomy

Guadalupe Gutiérrez-Soto, Iosvany López-Sandin, and Roberto Parra Saldivar

Highlights
Basidiomycetes are a promising group of fungi for upcycling agro-waste. Using basidiomycetes for agro-waste valorization contributes to a circular bioeconomy.

1 Introduction

The linear economic model presents significant global challenges such as depletion of natural resources, environmental degradation, and waste accumulation, characterized by a "take-make-dispose" pattern, which is unsustainable in the long term [1]. The bioeconomy offers an alternative paradigm aiming to transition from a linear to a circular model of production and consumption, utilizing biological resources like basidiomycetes in the valorization of agricultural residues to create a more sustainable and regenerative economic system [2]. Research related to the bioeconomy in Mexico has shown significant progress and potential for further improvement, particularly in utilizing agricultural residues for creating bio-businesses as a pathway for sustainable economic development [3]. Basidiomycetes use has emerged as a critical strategy in the bioeconomy, offering significant potential for valorizing biological resources and promoting sustainable practices [4]. Mexico boasts a vast diversity of fungi, including basidiomycetes, which are being

G. Gutiérrez-Soto (✉) · I. López-Sandin · R. P. Saldivar
Biomolecular Innovation Group, Laboratorio de Ciencias Naturales, Facultad de Agronomía, Universidad Autónoma de Nuevo León, Monterrey, Mexico

A. Meneses-Jácome et al. (eds.), *The Circular Bioeconomy in Industry*, Waste as a Resource, https://doi.org/10.1007/978-3-031-84359-4_4

studied for biotechnological applications [5]. Particularly in myco-biorefineries, research has focused on utilizing agri-food waste for enzyme production, animal feed supplements, and recovery of bioactive compounds [6, 7]. Recent studies have explored the potential of various fungal strains, such as *Pynoporus sanguineus* CS2 and *Trametes maxima* CU1, in diverse applications [6]. As a result, ongoing research is focused on exploring the potential of *Ganoderma resinaceum* CS27 in myco-biorefinery processes.

2 Material and Methods

The biological material consisted of the native strain *Ganoderma resinaceum* CS27, grown on agave bagasse (AB), orange peel (OP), and nutshell (NS), which were used as substrates. Two fermentation conditions were used: solid-state fermentation (SSF) and semisolid fermentation (SmSF). The laccase activity in the supernatants was determined using the method described by Abadulla et al. (2000), while cellulase and xylanase activities were measured following the protocol established by Miller (1959). Antioxidant capacity was estimated by the 1,1-Diphenyl-2-picrylhydrazine (DPPH•) radical inhibition. The fiber prebiotic activity was calculated using a mixed culture of *Lactobacillus acidophilus*, *Bifidobacterium lactis*, and *Lactobacillus casei* in an DeMan, Rogosa and Sharpe (MRS) media, supplemented with 10% fiber recovered from the spent substrates. The optical deviation was measured at 600 nm at 0 and 24 h using the McFarland scale. The results obtained were analyzed using the InfoStat statistical software. The minimum significant difference was determined through analysis of variance (ANOVA, $p \leq 0.05$). The results were expressed as mean values ± SD. For the mean analysis, a Tukey test was used at a significance level of 0.05 ($p \leq 0.05$).

3 Results and Discussion

The maxima laccase activity was detected in the NS medium under SmSF conditions (379 U/l), followed by the AB medium under SSF conditions (375 U/l) at 30 and 5 culture days, respectively (Fig. 1). To determine the presence of different isoforms due to the composition of the medium, an electrophoretic analysis is necessary. It has been reported that agricultural waste in the environment can significantly impact the production of laccase, an enzyme with various biotechnological applications. Additionally, it has been reported that utilizing agricultural waste products as carbon sources can enhance laccase production economically [8].

In the OP medium, no levels of this enzyme were recorded. In the cellulases case, the AB and OP media in SmSF showed higher levels (25 U/l) at 15 and 3 culture days, respectively (Fig. 2a). Also, the major xylanase levels at 3 (85 U/l) and 15 (48 U/l) culture days were observed in the OP medium (Fig. 2b). Basidiomycetes

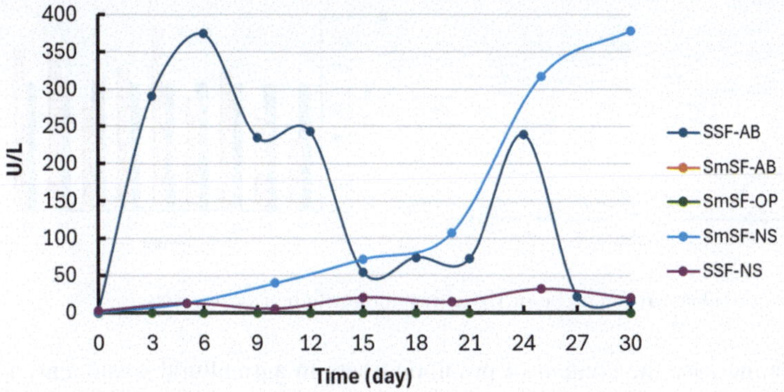

Fig. 1 Laccase production in agro-wastes

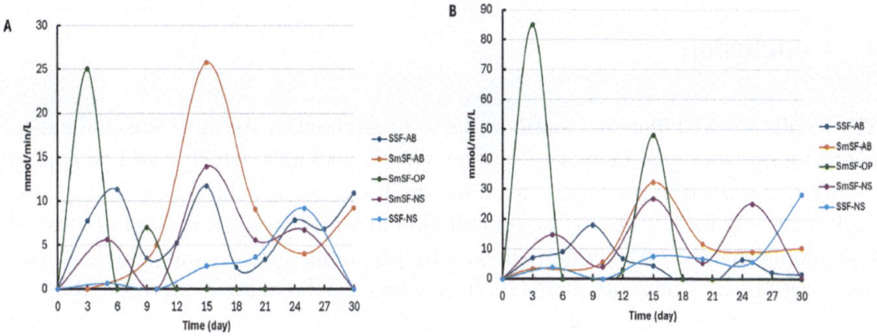

Fig. 2 Cellulase and xylanase production in agro-wastes

cultivated in agro-industrial waste have shown promising results in producing lignocellulolytic enzymes. These fungi possess a diverse enzymatic arsenal, including laccases, cellulases, and xylanases, crucial for lignocellulose degradation [6].

Additionally, the pectinase production curve in the OP medium was determined (data not shown), revealing that maximum activity (301 U/l) was reached after 9 days of culture. Regarding antioxidant activity, higher activity was detected in the OP medium. At time zero, 60% inhibition of the DPPH$^{\bullet}$ radical was quantified. However, this was increased over the cultivation period, with higher activity at 12 culture days (68%) (Fig. 3a). These findings highlight the potential of utilizing orange peel as a substrate for basidiomycete antioxidant production, emphasizing its role in enhancing enzymatic and nonenzymatic antioxidant activities in fungal systems [9].

Finally, the fiber's prebiotic activity showed that the OP medium condition increased the Colony Forming Unit (CFU) number, showing higher activity at 20 culture days (Fig. 3b). These results can be explained by the enzymatic activity of basidiomycetes, which act on the structural fibers of the plant cell wall. These

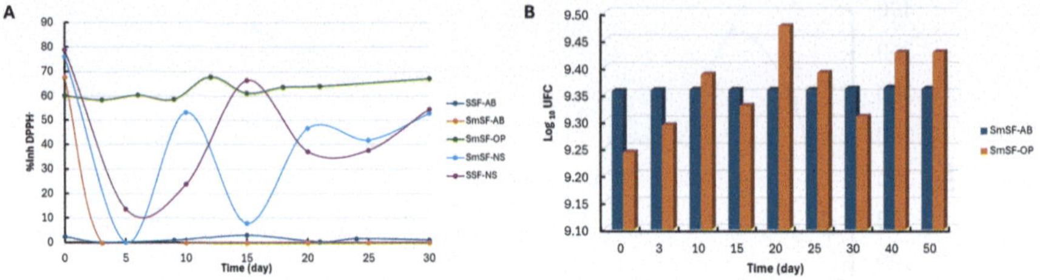

Fig. 3 Antioxidant (**a**) and prebiotic (**b**) activity production in agro-wastes

bacteria increase the content of prebiotic fibers in agricultural waste, enriching its nutritional and functional value. This allows the obtaining of higher-quality raw materials to produce functional foods and prebiotic supplements [10].

4 Conclusion

The results showed that enzymatic levels were affected by the agro-waste chemical composition, allowing the recovery of enzymatic cocktails, which could be applied in different industrial processes, from the modification of baking dough properties (xylanases and laccase) to the bioremediation of water bodies (laccase). It is worth mentioning that the OP medium allowed for obtaining supernatants with high pectinase activity and antioxidant metabolites, which can be of great interest in the food industry. At the same time, the fibers recovered after fermentation can be considered raw materials for the development of bio-functional foods, thanks to the probiotic effect shown in this study.

References

1. Keswani C (2020) Bioeconomy for sustainable development. Springer, Berlin
2. Kardung M et al (2021) Sustain For 13(1):413
3. Secretaría de Agricultura y Desarrollo Rural. https://www.gob.mx/agricultura/prensa/impulsa-sader-generacion-de-valor-agregado-en-productos-del-campo-con-el-desarrollo-de-la-bioeconomia#:~:text=La%20bioeconom%C3%ADa%20es%20una%20alternativa,la%20innovaci%C3%B3n%20que%20M%C3%A9xico%20necesita. consulted in Marzo 2024 (In Spanish)
4. Agyare C (2019) Bioactive metabolites from basidiomycetes. In: Advances in macrofungi: diversity, ecology and biotechnology. Taylor and Francis Group, Florida, pp 230–253
5. Aguirre-Acosta et al (2014) Biodiversidad de hongos en México. Revista mexicana de biodiversidad P 85:76–81. (In Spanish)
6. Gutiérrez-Soto G et al (2015) Bioresources 10(4):6676–6689
7. Hernández-Martínez CA et al (2022) Microbiol Res 13(4):1018–1026
8. González JC et al (2013) PLoS One 8(9):e73721
9. Lung MY et al (2010) J Food Sci 75(1):E18–E24
10. Pilafidis S et al (2022) Appl Sci 12(22):11426

Extracellular Compounds of Arthrospira platensis Under L-Tryptophan Supplementation: UHPLC Coupled with Mass Spectrometry (ESI-QTOF-MS) Identification

Natali López-Mejía, Ana Lucía Rangel Luna, Geison Modesti Costa, Hugo Alexander Martínez Correa, and Hugo Fabian Lobatón García

Highlights

- Arthrospira platensis showcased its capacity to generate extracellular pigments (neurosporaxanthin and pyropheophorbide a) when cultivated in a medium supplemented with L-tryptophan.
- In a medium supplemented with L-tryptophan, A. platensis demonstrated its ability to produce extracellular auxins—specifically indole-3 butyric acid and indole 3-propionic acid, both of which hold promise for incorporation into biostimulant formulations.

1 Introduction

Climate change and increasing food demand have spurred the excessive use of synthetic agro-inputs, worsening pollution, acid rain, and droughts while endangering biodiversity, human health, and food security. In response, bio-stimulants with

N. López-Mejía · H. A. Martínez Correa
Departamento de Ingeniería, Facultad de Ingeniería y Administración, Universidad Nacional de Colombia Sede Palmira, Palmira, Colombia
e-mail: nalopezme@unal.edu.co; hamartinezco@unal.edu.co

A. L. Rangel Luna · G. M. Costa
Grupo de Investigación en Fitoquímica, Departamento de Química, Facultad de Ciencias, Pontificia Universidad Javeriana, Bogotá, Colombia
e-mail: anarangel@javeriana.edu.co; modesticosta.g@javeriana.edu.co

H. F. Lobatón García (✉)
Vicerrectoría de Investigaciones, Universitaria Agustiniana, Bogotá D.C., Colombia
e-mail: hugo.lobaton@uniagustiniana.edu.co

A. Meneses-Jácome et al. (eds.), *The Circular Bioeconomy in Industry*, Waste as a Resource, https://doi.org/10.1007/978-3-031-84359-4_5

phytohormones are increasingly used to enhance fruit quality, nutrient absorption, and pest resistance. Among these phytohormones, auxins like Indole-3-Acetic Acid (IAA) are actively studied for their benefits in crops. Cyanobacteria, such as *Arthrospira platensis*, produce endogenous auxins like IAA and indole-3-butyric acid (IBA) (Zapata et al., 2021). Moreover, they contribute to the exogenous production of auxins in the supernatant, particularly IAA, via the tryptophan-dependent pathway, with amino acid concentrations ranging from 0.25 to 1.5 g L^{-1} [1]. *A. platensis* is capable of accumulating intracellular carotenoids in concentrations ranging from 0.38 to 4.43 mg g^{-1} [2]. These carotenoids, as characterized by Hynstova et al. [3], include zeaxanthin, lutein, astaxanthin, canthaxanthin, and β-carotene, with zeaxanthin being the predominant type. However, there is limited literature on the extracellular production of carotenoids by microalgae or cyanobacteria, with only two recent studies available. One of these studies is a review by Cheng et al. [4], which mentions the extracellular production of unique carotenoids, specifically keto carotenoid equinenone, in *Botryococcus braunii,* although the referenced study does not provide clear details on this aspect. Finally, Carvalho et al. [5] conducted research demonstrating that *Parachlorella kessleri,* when subjected to stress from high CO_2 concentrations, produced carotenoids extracellularly, although the specific types of carotenoids could not be identified. Despite the potential of supernatant, its utilization in biorefinery models is restricted due to challenges in separation and purification. However, using supernatant offers advantages as metabolites are naturally suspended, negating the need for extraction techniques. The aim of this research was to characterize supernatant metabolites, specifically auxins and carotenoids, in *A. platensis* cultivation under L-tryptophan supplementation using liquid chromatography coupled with mass spectrometry (ESI-QTOF-MS). This biorefinery approach yields valuable pigments and compounds using both biomass and supernatant.

2 Material and Methods

The experiments were conducted in three 1-L Erlenmeyer flasks, each containing 300 mL of culture with an initial optical density (OD) of 0.40 ± 0.1 of *Arthrospira platensis.* The flasks were placed in an incubation chamber and shaken in a shaker at 140 rpm at a constant temperature and light intensity. A LED light lamp (Atreum Lighting, Hydra-1000, China) was used as the light source. To study the effect of L-tryptophan concentration in the medium, 1 g L^{-1} of L-tryptophan was added to the Zarrouk medium and was maintained at a constant temperature of 33 ± 0.5 °C under a continuous light intensity of 20 μmol m^{-2} s^{-1} (measured on the walls of the flask) for 7 days until reaching an optical density (OD) of 0.85 ± 0.05. In liquid–liquid extraction, 50 mL of the supernatant, obtained after centrifugation and removal of biomass, was collected. The pH of the solution was adjusted to 2.8 by adding 50% (w/w) citric acid. The mixture was then transferred to a separation funnel, followed by the addition of 100 mL of analytical-grade ethyl acetate. After agitating the solution three times, it was allowed to settle for 5 min, resulting in the formation of two distinct phases, with the upper phase containing the ethyl acetate. The upper phase was carefully separated and subjected to rotary evaporation at 37 °C until a paste-like consistency was obtained. This paste was subsequently resuspended using 1 mL of 98% methanol. The resulting sample was stored in Eppendorf tubes covered with aluminum foil at 4 °C until it was ready for analysis.

The study utilized carotenoids and auxin dissolved in methanol, filtered through a 0.22-μm filter before injection. Qualitative verification was conducted using a Shimadzu Nexera LCMS-9030 UHPLC-ESI-QTOF-MS system. The Ultra High Performance Liquid Chromatography (UHPLC) setup comprised a gradient pump, autosampler, degasser, communication bus module, and column oven. Separation was achieved using a C18 Kinetex Core-shell UHPLC Column at 40 °C with a flow rate of 0.4 mL min^{-1} and an injection volume of 5 μL. The mobile phase consisted of acetonitrile and water with 0.05% acetic acid, with a specific gradient over time. Detection was carried out by a Photodiode Array (PDA) detector at various wavelengths. The QTOF-MS system operated in positive and negative modes with specified parameters including gas flow rates, temperatures, interface voltages, and mass range. The LC–MS/MS data were collected and processed using MZmine 2.53.

3 Results and Discussion

The UHPLC analysis resulted in two chromatograms obtained at nonvisible UV wavelengths (280 and 320 nm), where five main peaks were detected (Fig. 1a, b). The application of mass spectrometry with negative ionization confirmed that peaks 1–3 and 5 correspond to indole compounds. Specifically, the first peak corresponds to indole propionic acid (m/z 188.0367), the second to indole-3-aldehyde (IAld)

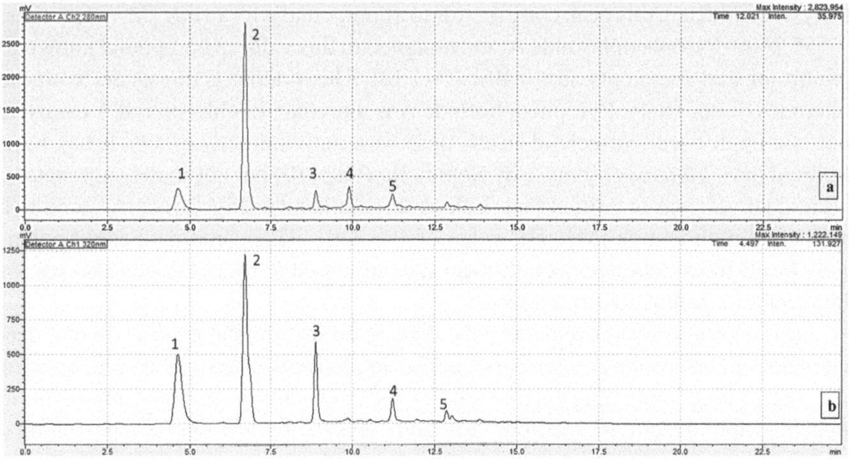

Fig. 1 Chromatograms obtained by UHPLC at 280 nm (**a**) and 320 nm (**b**)

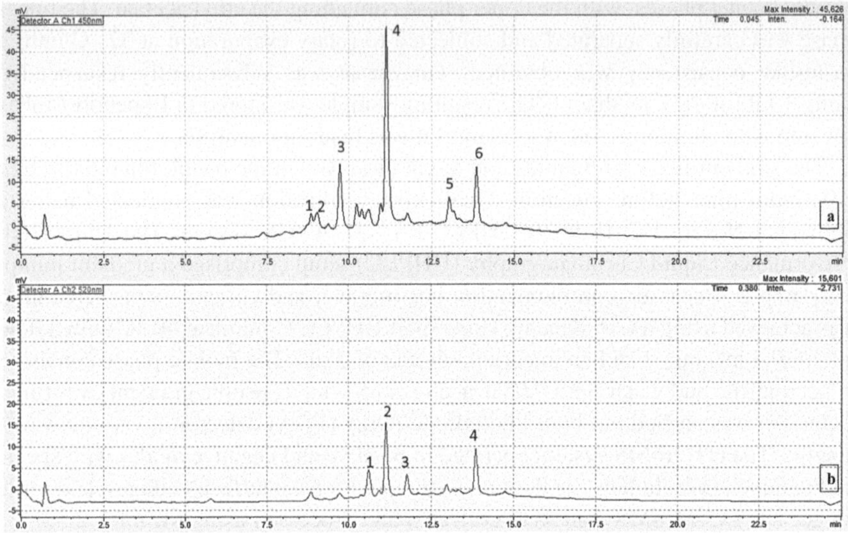

Fig. 2 Chromatograms obtained by UHPLC at 450 nm (**a**) and 520 nm (**b**)

(m/z 144.0455), the third to indole-3-butyric acid (m/z 202.0501), and the fifth to Indole-3-acetyl-tryptophan methyl ester (IATrp-Me) (m/z 374.151) [6, 7]. All the identified fragments of indole-type metabolites exhibited fragments with m/z $[M\text{-}H]^-$ values of 116.04, characteristic of these compounds [7]. Among these compounds, indole propionic acid and indole-3-butyric acid have been previously reported in *A. platensis* biomass (Zapata et al., 2021) and other cyanobacteria [6]. Furthermore, the presence of pigments was verified by the prominent peaks observed at 450 and 520 nm (Fig. 2a, b). The most intense peaks, peak 4 (Fig. 2a) and peak 2 (Fig. 2b), were obtained at a retention time (Rt) of 11.16. According to the mass analysis conducted, this peak corresponds to a compound with an m/z [M + H] + value of 535.2609. Pyropheophorbide A was assigned to this value, a compound known to exhibit ionized fragments at m/z 461.2341 [8]. This finding is consistent with the outcomes of our study. Pyropheophorbide A is a product of chlorophyll A catalysis and, although not a carotenoid itself, possesses antiviral activity [9]. It has been reported in *A. platensis* biomass by Sommella et al. [10] but not in the supernatant. Additionally, it was identified that peaks 6 (Fig. 2a) and 4 (Fig. 2b) correspond to a compound with an m/z $[M + H]^+$ value of 499.3491. This compound could potentially be all-trans-neurosporaxanthin, an apocarotenoid terpenoid that is also part of the metabolic pathway of *A. platensis*.

While significant work remains to be done in optimizing the production of external metabolites through *A. platensis*, such as implementing fed-batch cultivation or pH control and conducting further studies on the excretion mechanisms of these molecules into the culture medium, this study lays the foundation for envisioning a biorefinery process in which both the biomass and supernatant can be effectively utilized.

4 Conclusions

This study has demonstrated that, under the specific examined culture conditions involving the supplementation of a medium with L-tryptophan, it is indeed feasible to cultivate a medium characterized by high concentrations of auxins, including indole-3 butyric acid and indole-3-propionic acid. Importantly, this cultivation regime stimulates the extracellular biosynthesis of a pigment known as pyropheophorbide a, along with the carotenoid neurosporaxanthin. Significantly, this marks the pioneering documentation of extracellular pigment production, encompassing a carotenoid, by *A. platensis.* These findings lay the groundwork for the potential adoption of a biorefinery framework, facilitating the comprehensive utilization and exploitation of both the biomass and the supernatant components.

Compounds	Rt (min)	Elemental composition	Theoretical mass[1]	Observed mass (m/z)	MS[2]
Indole propionic acid	4.71	$C_{11}H_{11}NO_2$	189.0789	188.0367 (−3.19)	116.0494; 143.0360; 217.9448; 232.0232; 267.9644
Indole-3-aldehyde (IAld)	6.76	C_9H_7NO	145.0527	144.0492 (−24.29)	116.0494; 144.0445;187.0511
Indole-3 butyric acid	8.90	$C_{12}H_{13}NO_2$	203.0946	202.0898 (−23.75)	202.0501; 115.0427; 259.0855; 142.0288; 174.0554; 159.0316; 100.9038
Indole-3-acetyl tryptophan Methyl ester (IATrp-me)	11.24	$C_{22}H_{21}N_3O_3$	375.156	374.1655 (−29.13)	273.1010; 245.1082; 199.0494; 116.0494; 144.0445; 360.0966
Neurosporaxanthin; All-trans-neurosporaxanthin	13.75	$C_{35}H_{46}O_2$	498.3497	499.3208 (−57.87)	499.1922; 497.1776; 382.1317; 471.1683
Pyropheophorbide A	11.35	$C_{33}H_{34}N4O_3$	534.2625	535.2482 (−27.65)	461.1481; 517.1650; 503.1514

References

1. Ahmed M (2010) J Microbiol Biotechnol 20(9):1259–1265
2. Park W et al (2018) Molecules 23(8):2065
3. Hynstova V et al (2018) J Pharm Biomed Anal 148:108–118
4. Cheng P et al (2019) Bioresour Technol 291:121911
5. Carvalho P et al (2021) J Biotechnol 329:151–159
6. Duong T et al (2021) Chemosphere 284:131242
7. Revelou P et al (2019) Molecules 24(14):2615
8. Juin C et al (2015) Mar Drugs 13(4):2541–2558
9. Ratnoglik S et al (2014) Microbiol Immunol 58(3):188–194
10. Sommella E et al (2018) Molecules 23(5):1152

Study of Agro-Industrial Residues from Rice and Oil Palm from the Southeast Region of Colombia: Characterization, Annual Waste Generation, and Energy Potential

Sergio Andrés Morales Restrepo, Luis Sebastián Mendoza Castellanos, Ana Lisbeth Galindo Noguera, and Karina Angélica Ojeda Delgado

Highlights

- The chemical composition of agro-industrial residues from rice and oil palm was studied. The obtained results were compared with other studies available in the literature.
- A methodology was applied to determine the annual waste generation and the energy potential of lignocellulosic biomass.

1 Introduction

Agro-industrial residues, also known as lignocellulosic biomass, are produced during the harvesting and processing of agricultural crops. These residues are composed of cellulose, which is a major carbon source, hemicellulose, and lignin. Colombia has great potential for the utilization of lignocellulosic biomass as an energy source, especially in the southeast region. In 2021, approximately 194,879

S. A. Morales Restrepo (✉) · L. S. Mendoza Castellanos
Research Group in Resources, Energy and Sustainability (GIRES), Faculty of Energy Engineering, Universidad Autónoma de Bucaramanga (UNAB), Bucaramanga, Colombia
e-mail: smorales@unab.edu.co; lmendoza630@unab.edu.co

A. L. Galindo Noguera
Research Group in Resources, Energy and Sustainability (GIRES), Faculty of Energy Engineering, Universidad Autónoma de Bucaramanga (UNAB), Bucaramanga, Colombia
e-mail: agalindo144@unab.edu.co

K. A. Ojeda Delgado
Process Design and Biomass Utilization Research Group (IDAB), Chemical Engineering Department, Universidad de Cartagena, Cartagena, Colombia
e-mail: kojedad@unicartagena.edu.co

A. Meneses-Jácome et al. (eds.), *The Circular Bioeconomy in Industry*, Waste as a Resource, https://doi.org/10.1007/978-3-031-84359-4_6

and 78,863 hectares were planted with rice and oil palm crops, respectively, in this region [1]. As a result, significant amounts of residues are expected to be generated, which are utilized in specific applications. Rice husk is primarily used in the poultry sector, while palm kernel shells, mesocarp fiber, and empty fruit branches serve as boiler fuel or fertilizer in plantations [2, 3]. However, these applications do not represent optimal valorization since they do not fully harness the true potential of biomass. This occurs because the residues do not add value to the original process that generated them, and sometimes both companies and the community are unaware of the potential they possess.

In this sense, it is crucial to understand the chemical and structural composition of the lignocellulosic biomass generated from rice and palm crops in the southeast region of Colombia. This understanding will provide valuable support for decision-making regarding valorization pathways and future biorefinery schemes. Fast proximate analysis was obtained by thermogravimetric analysis (TGA) methodology. The results were compared with literature data and showed good agreement, taking into account that biomass properties and characteristics can change from one region to another. These findings can play a crucial role in expanding the range of valorization options for residues from the rice and palm oil sectors. By leveraging this information, it becomes possible to improve process efficiency, optimize resource utilization, and further enhance sustainability outcomes.

2 Material and Methods

2.1 Materials

The materials used in this research are samples of rice husk, palm kernel shell, empty fruit branch, and mesocarp fiber, all provided by agro-industrial companies from Casanare (5°26′0″ N, 71°30′0″ W), located in the southeast region of Colombia.

2.2 Methods

Proximate analysis by TGA was conducted using a TGA analyzer (TA Instrument TGA Q500) with samples of 10 mg average of each biomass. This technique measures the loss or gain of weight of a substance as a function of heating at high temperatures (from ambient to 900 °C) and can also be used to determine the heating value [4]. The weight loss occurs due to processes such as evaporation, sublimation, decomposition, desorption, reduction, oxidation, and absorption. This device applies a heat ramp of 10.00 °C/min to 700.00 °C.

On the other hand, annual waste generated by these crops (as received) in the Department of Casanare was estimated using information available in public databases and the following equation [1, 5]:

$$\mathrm{Aw} : A^{*}\mathrm{Cy}^{*}\mathrm{Rrp} \tag{1}$$

where

Aw: Annual residue generation potential (t/yr)
A: Planted area of the crop in the region (ha)
Cy: Crop yield (tons of product/cultivated area)
Rrp: Relation residue/product (tons of residue/tons of principal product)

Based on the data obtained from Eq. 1 and knowing the low heating value (LHV), the annual energy potential of agricultural residual biomass can be calculated. This may include estimating the amount of energy that can be obtained through direct combustion, biogas production, or bioethanol generation, among other processes, thereby contributing to the diversification of energy sources and the reduction of greenhouse gas emissions.

3 Results and Discussion

Table 1 shows a proximate analysis of biomass samples of rice husk obtained by TGA and its respective comparison with other studies available in the literature.

According to the information in Table 1, proximate analyses of rice husk from different authors show small differences in some characteristics, such as moisture content and volatile matter. Other characteristics, such as fixed carbon and ash content, show greater variation. This can be attributed to specific characteristics of the region in which the sample was obtained and test conditions. Table 2 shows a proximate analysis of biomass samples of palm kernel shell, mesocarp fiber, and empty fruit branch obtained by TGA and its respective comparison with other studies available in the literature. According to this information, the obtained results of proximate analysis of oil palm biomass are very similar in both, the present study and literature results. Some small differences can be seen because these biomass samples are from different regions. In addition, oil palm biomass showed higher volatile matter content than rice husk, which has a significant impact on the

Table 1 Proximate analysis of rice husk from Casanare, Colombia, and comparison with other studies

	Moisture (%)	Volatile matter (%)	Fixed carbon (%)	Ash (%)
Present study	6.403	56.1	7.149	30.33
Maksum et al. [6]	8.70	54.10	13.35	23.85
Nisha et al. [7]	9.8	64	8.8	17.4
Kuo et al. [8]	7.14	70.60	8.34	13.93

Table 2 Proximate analysis of oil palm biomass from Casanare, Colombia by TGA method

Literature average results (Qureshi et al. [9])				
Biomass	Moisture content (%)	Volatile matter (%)	Fixed carbon (%)	Ash (%)
Empty fruit branch	6.9–9.9	55.9–83.9	8.65–19.9	1.6–14.2
Mesocarp fiber	4.7–8.6	66.8–78	7.6–24.4	3.5–10.2
Palm kernel shell	4.2–12	36.8–77.5	13.2–54.7	1–10.5
Obtained results				
Biomass	Moisture content (%)	Volatile matter (%)	Fixed carbon (%)	Ash (%)
Empty fruit branch	2.30	69.12	14.67	13.91
Mesocarp fiber	6.35	68.85	5.13	19.66
Palm kernel shell	3.40	62.41	13.81	20.38

Table 3 Annual waste generation from rice and oil palm crops in Casanare, Colombia

Biomass	Annual waste generation (tons/year)	LHV (MJ/ton) [5]	Annual energy potential (TJ/year)
Rice husk	227,613	15,085.04	3433.5
Palm kernel shell	58,469	16,697	976.25
Mesocarp fiber	167,434	17,894.4	2996.13
Empty fruit branch	281,714	16,835.12	4742.7
Total	735,230	–	12,148.58

efficiency and control of biomass conversion processes, such as gasification and pyrolysis. In gasification, for instance, the volatile material content can improve speed production and syngas quality. Therefore, these biomasses could be co-gasified to overcome the limitations of standard gasification processes.

Table 3 shows annual waste generation from rice and oil palm crops in Casanare, Colombia according to Eq. 1. The results show that there is a theoretical availability of 735,230 tons/year of biomass, which indicates a great potential for future studies about biorefinery schemes to obtain bioenergy and high-value products. Additionally, by knowing the LHV of rice husk and oil palm biomass from Colombia, the annual energy potential of these crops was estimated at 12,148.58 TJ/year for the Casanare region.

4 Conclusion

The proximate analysis of lignocellulosic biomass using the TGA method is a relatively low-cost and fast approach to determining the main characteristics of biomass samples. In this study, the results obtained from samples of rice and oil palm biomass were compared with data available in the literature, showing good agreement. However, it is important to note that there may be differences in the proximate analysis results of biomass samples between different studies due to specific characteristics that crops may exhibit in different regions. Estimation of annual waste generation and annual energy potential from these two crops in Casanare, Colombia, was calculated to be around 735,230 tons/year and 121,148.58 TJ/tons, respectively. These results can serve as valuable references for future studies focusing on the valorization of these residues in the region.

References

1. AGRONET Database. https://www.agronet.gov.co/estadistica—consulted in February 2024
2. Terry L et al (2021) Carbon Resour Convers 4:239–250
3. Ahmad F et al (2019) Renew Sust Energ Rev 109:386–411
4. Park S et al (2022) Energy Rep 8:12038–12044
5. Escalante H et al (2010) Energy potential of residual biomass in Colombia. UPME
6. Maksum A et al (2018) IOP Conf Ser Mater Sci Eng 316:012050
7. Das N et al (2021) Biomass Convers Biorefinery 13(2):1503–1527
8. Kuo L-A et al (2023) PRO 11(11):3119
9. Qureshi SS et al (2019) Biomass Convers Biorefinery 9(4):827–841

Development of Microbial Cellulose Dressings from Coffee Mucilage: A Matrix for Controlled Topical Drug Delivery

Ingrith Carolina Flórez García, Víctor Alfonso Solarte David, and Silvia Milena Becerra Bayona

Highlights

- The microbial cellulose was synthesized in situ using only coffee mucilage.
- Reusing coffee mucilage to cultivate microbial cellulose can reduce costs and improve the management of agricultural waste.
- By adjusting the culture medium, microbial cellulose hydrogels can be designed with specific physical properties for dressing design.

1 Introduction

Biomaterials, a captivating and interdisciplinary field, have become fundamental in enhancing conditions and quality of life in modern society. This use is particularly notable in the medical field, where biomaterials are primarily used and may be exposed, directly or indirectly, to biological systems, thus playing a crucial role in various therapeutic applications [1]. Biomaterials have been developed for a range of applications, from supporting bodily functions to drug administration and discovery [2]. Microbial cellulose, a biodegradable biomaterial produced by certain bacteria, has a rich history and diverse applications. It is mainly produced by acetic acid bacteria, particularly *Komagataeibacter xylinus*, and possesses unique properties such as high water retention capacity, mechanical strength, and biocompatibility [3, 4]. The use of microbial cellulose in dressings for chronic wounds, c drug *delivery* systems has been explored due to its biocompatibility and biodegradability [5].

I. C. Flórez García (✉) · V. A. S. David · S. M. Becerra Bayona
Universidad Autónoma de Bucaramanga—UNAB, Bucaramanga, Colombia
e-mail: iflorez167@unab.edu.co

A. Meneses-Jácome et al. (eds.), *The Circular Bioeconomy in Industry*, Waste as a Resource, https://doi.org/10.1007/978-3-031-84359-4_7

Despite its potential, the main challenge is to increase production, which has led to research into alternative production methods [6], specifically the production of microbial cellulose from agricultural waste materials.

Currently, the intense generation of biomass from agricultural products is causing environmental issues and incurs high economic costs associated with its disposal and regeneration [7]. One such agricultural product with potential for residual biomass is precisely coffee, where only 8% of the coffee cherry is converted into the beverage we enjoy, while the remaining 92% becomes by-products, leading to consider several important aspects of sustainability and efficiency in coffee production [8]. For this reason, it is necessary to search for methods to utilize by-products in other areas, such as raw material in the creation of derivative products. This low utilization rate also highlights the need to innovate processes that can increase the efficiency of the use of the coffee cherry. The challenge lies in finding sustainable alternatives that allow a circular economy in which all by-products are reused, reducing environmental impact, and adding value to the production chain. In this context, this research aims to produce microbial cellulose hydrogel dressings from coffee mucilage with the potential to be used in the future as a biomaterial for chronic wounds.

2 Material and Methods

In the study, the *Gluconacetobacter xylinus* strain ATCC 23768 was used for the production of microbial cellulose hydrogel dressings, using three different culture media: a standard Hestrin-Schramm medium (HS), a Hestrin-Schramm medium modified with coffee mucilage (HSCM), and a medium composed solely of coffee mucilage (CM). The strain was reactivated under aseptic conditions and incubated at 26 °C for 24 h. Subsequently, the cultures were cryopreserved by centrifugation and addition of Dimethylsulphoxide (DMSO), and stored at −80 °C to ensure their viability for 5–10 years. Aerobic fermentation was conducted at 30 °C, adjusting the pH of the media to 5.5 with NaOH and HCl solutions. Finally, the microbial cellulose membranes were purified using 0.5% NaClO and multiple washes until a neutral pH was reached, and they were stored at 4 °C for preservation.

3 Results and Discussion

The results obtained in the production of hydrogels based on microbial cellulose are presented below. In the experimental phase, once the inoculum was prepared, it was subjected to static fermentation. This stage aimed at creating hydrogels by day 7, using an inoculum concentration of 15% v/v and employing three different culture media: HS, HSCM, and CM. The D_{HSCM} and D_{CM} hydrogel dressings were cultivated over 5 days of fermentation, and the dressing from the standard medium D_{HS}

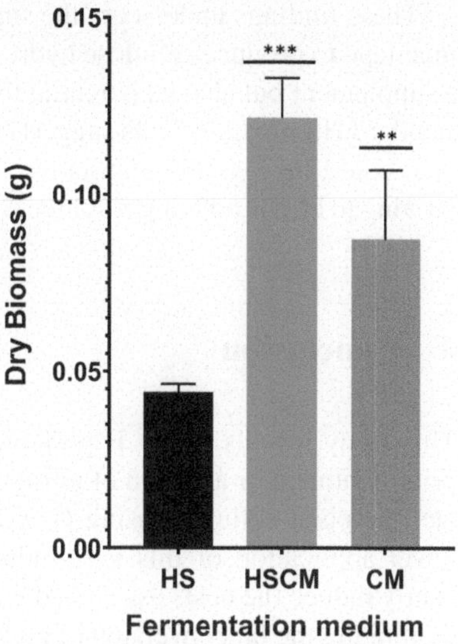

Fig. 1 Biomass of the microbial cellulose hydrogel dressings. **Note:** The dry biomass of the microbial cellulose hydrogels obtained for each of the fermentation media HS, HSCM, and CM is presented. Data are reported for $n = 6$ independent samples per formulation. Statistical differences between the media are indicated by *** for $p < 0.0001$ and * for $p = 0.0071$

was obtained at 7 days. Figure 1 shows the dry biomass of the microbial cellulose hydrogel dressings obtained for each of the fermentation media HS, HSCM, and CM.

The study on microbial cellulose hydrogels demonstrated that the composition of the culture medium is a determining factor in the resulting biomass. In particular, the hydrogels cultured in the HSCM medium accumulated a dry biomass three times greater than those cultured in the standard HS medium, a result that showed high statistical significance ($p < 0.0001$) and 2.3 times greater compared to the hydrogels developed in the CM medium, which contained only coffee mucilage, with also notable statistical significance ($p = 0.0071$). An increase in dry weight was observed among the fermentation media, with weights ranging from a minimum of HS: 0.04 ± 0.002, HSCM: 0.12 ± 0.011, and CM: 0.09 ± 0.019 g of microbial cellulose, while the yield varied from HS: 0.80, HSCM: 2.40, and CM: 1.80 g/l, respectively. The cellulose yield from the HS medium was the lowest compared to the other media. The productivity of microbial cellulose increased 2.3 times in the CM medium (0.25 g/l *day) compared to HS (0.11 g/l*day). The HSCM medium had the highest cellulose productivity (0.34 g/l*day).

In a report by Florea, M., they recorded a dry microbial cellulose biomass of 0.25 g obtained over 6 days of static fermentation [9], values close to those found. Meanwhile Hamed, D. A. (2023) managed to obtain a cellulose yield 1.7 times higher than the standard medium from black tea, achieving 3.06 ± 0.205 g/l in the black tea only medium and 1.8 ± 0.234 g/l in the HS medium [10]. Using only tobacco waste as a carbon source, they report a production of 1.8 g/l of microbial cellulose [11]. In a study by Huang, C. et al. (2016) only lipid wastewater was used, and the maximum cellulose yield reached was 0.66 g/l [12], slightly lower compared to the results of the present work where only coffee mucilage was used.

These findings underscore the superiority of culture media enriched with coffee mucilage to produce cellulose hydrogel dressings, indicating not only its efficacy as a supplement but also its potential to be used as the sole component, surpassing the standard HS medium. This suggests that coffee mucilage, as an agro-industrial residue, could be a viable alternative for large-scale microbial cellulose production, proving to be a promising resource for the accelerated growth of microbial cellulose.

4 Conclusion

This study reveals that it is possible to use coffee mucilage directly, without any prior treatment or addition of nitrogen or any other element, as a substrate to generate microbial cellulose using *G. xylinus* ATCC 23765. This approach proposes a novel application of this agro-industrial waste. This methodology could significantly reduce the costs associated with the production of microbial cellulose, while facilitating the bioconversion of glucosidic monomers into high-value carbohydrate biopolymers, such as those used in the manufacture of dressings that could be used on chronic wounds. The study demonstrated that an agricultural byproduct, specifically coffee mucilage, can be utilized to cultivate microbial cellulose using the bacterial strain *Gluconacetobacter xylinus* ATCC 23765. This approach presents an opportunity to reduce the production costs of MC by repurposing a residue that is naturally rich in glucose. This method not only offers a cost-effective alternative for generating value-added products but also addresses the challenge of disposing of agricultural waste, which can be phytotoxic. Furthermore, the microbial cellulose produced in this manner exhibited superior performance, as evidenced by the mechanical properties of the hydrogel dressings, which were more similar to those obtained from standard HS culture media.

The results show that by modifying the culture medium, it is possible to produce microbial cellulose hydrogels with varied physical properties, including significant differences in elasticity and compression moduli. This indicates that by adjusting these parameters, the hydrogels can be customized for specific applications such as dressing design, thereby taking advantage of their diversity in stiffness and compression resistance.

References

1. Kumar V et al (2023) Curr Mater Sci
2. Hassan S et al (2020) Biomaterials for on-chip organ systems. in
3. Gullo M et al (2018) Appl Microbiol Biotechnol 102:6885–6898
4. Popa L et al (2022) Materials (Basel) 15
5. Rajwade JM et al (2015) Appl Microbiol Biotechnol 99:2491–2511
6. Pineda L et al (2012) Ing y Cienc 8:307–335
7. Candra A et al (2024) J Drug Deliv Sci Technol 105477:105477

8. Valencia NR, Zambrano-Franco DA (2010) Los subproductos del café: fuente de energía renovable. Chinchiná, Colomb. Cenicafé—(In Spanish)
9. Florea M et al (2016) Proc Natl Acad Sci USA 113:E3431–E3440
10. Hamed DA et al (2023) World J Microbiol Biotechnol 39:48
11. Ye, J. et al. (2019) Bacterial cellulose production by Acetobacter xylinum ATCC 23767 using tobacco waste extract as culture medium. Bioresour. Technol. 518–524. https://doi.org/10.1016/j.biortech.2018.12.028
12. Huang C et al (2016) Carbohydr Polym 136:198–202

Extraction of Pectin from Citrus Wastes in Santander (Colombia)

Walter Pardavé Livia and Paula Mendoza Rojas

Highlights

- This short communication deals with the exploratory research to implement a project to produce pectin from citrus wastes as a potential contribution to the circular bioeconomy in the region of Santander (Colombia).
- The study demonstrates that it is possible to reduce solvent use by 50% without affecting the product yield.

1 Introduction

Pectin is a polysaccharide used as a natural additive in various food industries, serving as a gelling, thickening, stabilizing, and emulsifying agent for other products. Likewise, in the pharmaceutical industry and biotechnology area, pectin is used in the formulation of drugs, cosmetics, and therapeutic purposes, among others [1, 2]. Pectin is present in all plant tissues, although, due to its content, only some varieties are suitable for commercial extraction. Currently, at the industrial level, pectin is commonly extracted from apple or orange waste, derived from cider and juice

W. P. Livia (✉)
Instituto de Investigaciones XERIRA – UDES, Bucaramanga, Colombia
e-mail: wal.pardave@mail.udes.edu.co

P. M. Rojas
Instituto de Investigaciones XERIRA – UDES, Bucaramanga, Colombia

Universidad de Santander – UDES, Bucaramanga, Colombia

A. Meneses-Jácome et al. (eds.), *The Circular Bioeconomy in Industry*, Waste as a Resource, https://doi.org/10.1007/978-3-031-84359-4_8

production [3]. Other interesting alternative sources for the extraction of pectin are industrial wastes from the processing of papaya, mango, peach, sunflower, coffee, and cocoa [3–7]. Conventional pectin extraction requires the heating of high amounts of solvent to temperatures above 60 °C for periods typically longer than 30 min and up to 6 h [3], which represents a high energy consumption, in contrast to the current trend of using increasingly energy-efficient production methods in the industrial field [8, 9].

2 Material and Methods

2.1 Collection, Pretreatment, and Preservation of Raw Material

Approximately 8 kg of orange waste, including the albedo (or white inner part of the peel), seeds, and membranes, were collected from different juice vendors to have a sample with varied fruit characteristics and degrees of maturity. The samples were dried to 10% humidity, crushed, and stored protected from light and humidity (Fig. 1). The material resulting from the described procedure is referred to as raw material in the present investigation. By means of two pectin extraction operations (at the beginning and at the end of the study, 2 months later), under identical conditions, it was verified that the raw material did not degrade during the period of development of the study. Figure 2 shows the process of obtaining the sample.

Fig. 1 Citrus crops in Lebrija (Santander, Colombia)

Fig. 2 Collection of citrus peels in industry

2.2 *Variables of the Processes Involved*

Although there are several variants of the process for obtaining pectin, the following is a description of the basic and typical process, which is divided into three phases: pretreatment, extraction, and homogenization. The pretreatment consists of drying the material from which the pectin will be extracted to approximately 10% humidity, followed by crushing. Extraction basically consists of acid hydrolysis (the most important stage) under specific conditions of pH, temperature, and time, which are the independent variables that determine the quality of the product and the yield of the process.

The subsequent stages are as follows: the clarification operation, where the pectin solution is separated from the bagasse, alcoholic precipitation, and subsequent centrifugation or filtration, where the pectin is isolated from the solution. Finally, the wet pectin is dried with hot air to an appropriate humidity, and then the particle size of the product is homogenized by pulverizing as detailed in Fig. 3.

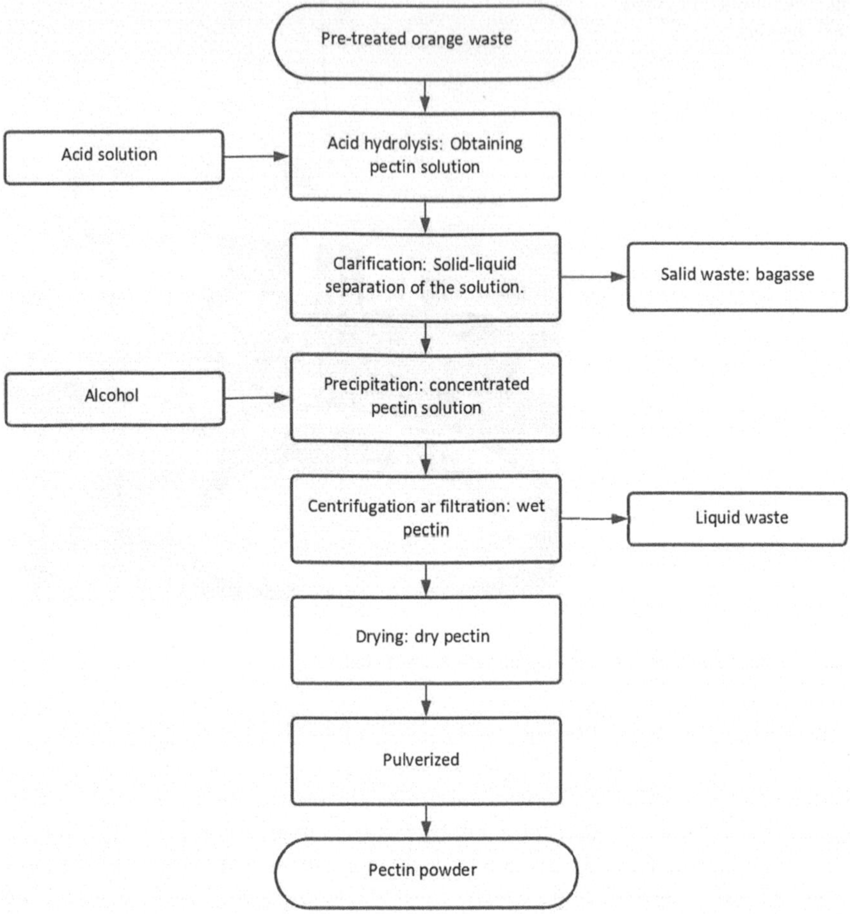

Fig. 3 Diagram of the stages in the process of obtaining pectin

3 Results and Discussion

3.1 Characterization of Citrus Peels

The moisture determined for citrus residues, comprising peel albedo, seeds, and fruit membranes, was 83.35%. The range of yield variation is wide (from 2.5% to 21.1%), which shows that at least one of the variables studied has a significant effect on the yield. The percentage of galacturonic acid fluctuates between 53.3% and 64.8%, which is below the limit of 65% established for commercial pectins.

Fig. 4 Steps to obtain pectin powder

3.2 *Product Obtained*

A sample of the pectin powder obtained in the laboratory by the microwave extraction method, using the optimum operating parameters, with a humidity of 8.7%, is shown in Fig. 4.

4 Conclusion

The following optimum values are established for the hydrolysis operation: Within the pH range 2.17–2.56, 2.17 is established as the optimum value because, with this value, a higher yield and a higher percentage of galacturonic acid are obtained. For a solvent:raw material ratio (mL:g) within the range of 18:1–24:1, no significant variation in yield or pectin quality was detected. Therefore, it is possible to use the minimum amount (18:1) of solvent without affecting these variables. For microwave extraction, periods between 7 and 15 min also had no significant effect on yield and product quality. Regarding the precipitation operation, it is concluded that it is feasible to reduce the percentage of solvent to 50% of the solution to be precipitated, without affecting the yield of the product. Lower values run the risk of not allowing precipitation, as occurred for a value of 40% solvent.

References

1. Espinal CF, Martínez HJ, Peña Y (2005) La cadena de cítricos en Colombia: Una mirada global de su estructura y dinámica 1991-2005. Ministerio de Agricultura y Desarrollo Rural, Observatorio Agrocadenas Colombia, Bogotá
2. Marín RF et al (2007) Food Chem 100(2):736–741
3. Wilkins MR et al (2007) Bioresour Technol 98(8):1596–1601

4. Kim WC, Lee DY, Lee CH, Kim CW (2004) J Food Eng 63(2):191–197
5. Instituto de Investigación de Recursos Biológicos "Alexander von Humboldt" (2003) Biocomercio sostenible. Estudio del mercado colombiano de aceites esenciales. Bogotá. (In Spanish)
6. Mesbahi G, Jamalian J, Farahnaky A (2005) Food Hydrocoll 19(4):731–738
7. Liu Y, Shi J, Langrish TAG (2006) Chem Eng J 120:203–209
8. Kingston HM, Jassie LB (2008) Introduction to microwave sample preparation, theory and practice. American Chemical Society, Washington, pp 7–31
9. Liu Y et al (2010) J Agric Food Chem 49(6):3051–3057

Influence of Different Growing Media on *Arthrospira platensis* Biomass Yield

Natalia Andrea Plata Ramírez, Jasmín Lorena Durán-Ortiz, and Alexander Meneses-Jácome

Highlights

- Arthrospira platensis was used in different culture media to evaluate biomass production
- The best biomass yield was achieved in the simulated seawater culture medium
- The 1:1 ratio of domestic wastewater to seawater as a culture medium enables a high biomass yield

1 Introduction

An estimated 359 billion cubic meters of wastewater is produced worldwide each year, of which 63% is collected, 52% is treated, and 48% is discharged into the environment without any treatment [1]. This leads to alterations in surface water bodies, which are impacted with nutrients such as N and P coming from activities such as agriculture, livestock, and fish farming, as well as industrial processes and urban effluents, which means that water bodies receiving these discharges become degraded, losing their self-purification capacity and becoming more susceptible to anoxia and eutrophication phenomena [2, 3].

N. A. Plata Ramírez (✉) · J. L. Durán-Ortiz · A. Meneses-Jácome
Circular Bioeconomy Research Center (ε-BiO): Biotechnology and Environment Group, Universidad Autónoma de Bucaramanga (UNAB), Bucaramanga, Colombia
e-mail: nplata@unab.edu.co

A. Meneses-Jácome et al. (eds.), *The Circular Bioeconomy in Industry*, Waste as a Resource, https://doi.org/10.1007/978-3-031-84359-4_9

It has been evidenced that microalgae and cyanobacteria cultures have been used for some decades as treatments for nutrient-polluted waters, especially those of municipal type [4]. This practice presents a significant opportunity in coastal areas, where high-yielding marine algae species can be used, and the mixture of effluents with seawater as substrate could be suggested to enhance the sanitation system [5]. Indeed, seawater/wastewater blending is a widely studied strategy for coastal regions requiring the remediation of different types of effluents. This approach not only contributes to sanitation but also increases the availability of biomass with commercial potential. For instance, microalgae have been applied as a complementary treatment of municipal wastewater for its reuse in productive activities of semiarid coastal regions of Brazil [6]. In addition, more recent studies have shown that microalgae and cyanobacteria such as *Dunaliella* sp, *Nannochloropsis salina*, and *Arthrospira platensis* are able to grow in saline media, when mixed with municipal wastewater, evidencing a viable and attractive option for domestic wastewater treatment in coastal areas [7]. Based on these evidences, this exploratory study has been driven, with the purpose of evaluating the biomass production of *Arthrospira platensis* in four different synthetic culture media: Zarrouk medium (ZM), synthetic domestic wastewater (DW), synthetic seawater (SW) and the mix DW-SW.

2 Material and Methods

This study has evaluated the growth and biomass yield of *Arthrospira platensis* sp., a species whose adaptability for this kind of application has been previously reported [7] and could be thought of as a reference to develop the treatment of domestic effluents mixed with seawater in the Colombian Caribbean Region. Physicochemical characterizations of human effluents and seawater from the "La Guajira" coast in Colombia were taken as a reference to prepare the synthetic DW and SW [8]. Tests were conducted in different culture media such as Zarrouk standard medium, synthetic DW, and synthetic SW, where it was shown that *Arthrospira platensis* grows well and survives long periods. The algal biomass yield was followed indirectly by the determination of TSS taking as reference method 2540D of Standard Methods for the Examination of Water and Wastewater (23rd Edition), and its relationship with pH, electrical conductivity (EC), and salinity was established. Figure 1 shows the initial set-up with the four culture media evaluated.

3 Results and Discussion

The biomass yield is drawn in Fig. 2. Here, the culture in ZM, which has been taken as the reference medium has the highest production of TSS, reaching w 488 mg/L after 30 days, followed by SW, DW-SW, and DW with values of 352, 274, and 98

Fig. 1 Culture media of *Arthrospira platensis* (from left to right, by pairs of jars: cultivated in ZM, DW, SW, and DW: SW mix, respectively)

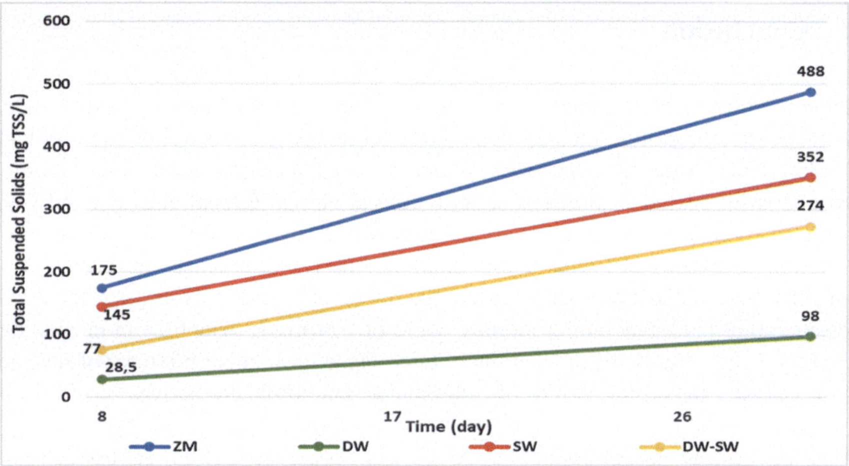

Fig. 2 Biomass yield of *Arthrospira platensis* sp. in different culture media

mg/L, respectively. It is highlighted that the DW:SW mixture in a 1:1 ratio proved to be a very effective culture medium for *Arthrospira platensis* species, exhibiting a promising biomass yield. On the other hand, Fig. 3 shows the pH evolution in the four culture media. The pH increases in all culture media without significant changes in the media salinity. The pH reaches an intermediate end-value for the culture raised in the DW:SW mixture regarding the DW and SW cultures. This behavior could be associated with nutrient removal processes and CO_2 metabolization by the microalgae [9, 10].

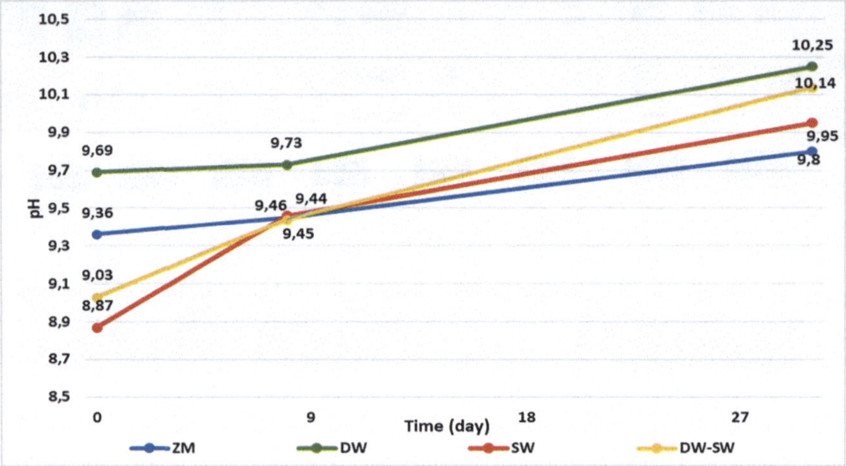

Fig. 3 Biomass yield of *Arthrospira platensis sp.* in different culture media

4 Conclusion

Three alternative culture media were contrasted against the ZM used as a reference, showing that the culture medium with the best biomass yield was SW and DW had the lower performance. *Arthrospira platensis* has demonstrated the ability to survive between 12 and 37 practical salinity units (PSU) of Zarrouk medium and SW, respectively, which reveals a remarkable ability to adapt to a wide range of salinity. Likewise, a mixture of synthetic DW and SW, in a 1:1 volumetric ratio, was determined to be a viable culture medium for the *Arthrospira platensis* sp., as reported in the literature [11], reaching a biomass yield of 274 mg/L, which is an interesting starting point to study the application of microalgae for treating mixtures of sewage with seawater in coastal regions of Colombia as sanitation alternative.

Acknowledgments Special thanks to the Ministry of Science, Technology, and Innovation of Colombia (MinCiencias) for the "Bicentenario" doctoral scholarship awarded to Ms. Natalia Plata for the period 2021–2025.

References

1. Jones ER et al (2021) Earth Syst Sci Data 13(2):237–254
2. Nogueira SMS et al (2018) Rev Ciênc Agron 49(4):599–606
3. Jiao W, Q. et al (2013) Ecol Indic 34:356–360
4. Pittman JK et al (2011) Bioresour Technol 102(1):17–25
5. Yang J et al (2010) Bioresour Technol 102(1):159–165
6. Marangon BB et al (2020) J Water Process Eng 37:101483
7. Brown JJ et al (2018) Sustainability 10(5):1–16

8. Estrada E et al (2020) Universidad de La Guajira. Riohacha, Colombia, pp 25–43. (In Spanish)
9. García-Gozalbes CC et al (2015) Tecnol Cien Agua 1:49–68
10. Holanda M et al (2022) J World Aquacult Soc 53(3):669–680
11. Zhou W et al (2017) Bioresour Technol 245:10–17

Technological Roadmap for Sustainable Hydrogen Production from Lignocellulosic Biomass Gasification for Energy Applications: A Perspective

Katerine Angélica Serrano, Yurley Paola Villabona Durán, Karina Angelica Ojeda Delgado, Ciro Eduardo Rozo Correa, and Luis Sebastián Mendoza Castellanos

Highlights

- Proposal of a technological roadmap for hydrogen production from lignocellulosic biomass.
- Identification of fluidized-bed gasification as the optimal technology for hydrogen production.
- Evaluation based on energy efficiency, environmental impact, and economic feasibility.

1 Introduction

The escalating demand for renewable energy solutions underscores the critical role of innovative technologies in addressing climate change and achieving sustainability goals. Hydrogen, as a clean energy carrier, offers significant promise due to its high energy yield and zero-emission profile when combusted. Lignocellulosic

K. A. Serrano (✉) · Y. P. Villabona Durán · C. E. R. Correa
Universidad Santo Tomás – Bucaramanga, Bucaramanga, Colombia
e-mail: katerine.acevedo01@ustabuca.edu.co; yurley.villabona@ustabuca.edu.co; ciro.rozo@ustabuca.edu.co

K. A. Ojeda Delgado
Universidad de Cartagena, Cartagena, Colombia
e-mail: kojedad@unicartagena.edu.co

L. S. Mendoza Castellanos
Universidad Autónoma de Bucaramanga – UNAB, Bucaramanga, Colombia
e-mail: lmendoza630@unab.edu.co

A. Meneses-Jácome et al. (eds.), *The Circular Bioeconomy in Industry*, Waste as a Resource, https://doi.org/10.1007/978-3-031-84359-4_10

biomass, an abundant and renewable resource, has emerged as a viable feedstock for sustainable hydrogen production through advanced gasification technologies. This study examines the technological viability and environmental impacts of various biomass gasification methods, with a particular focus on optimizing energy efficiency and minimizing ecological footprints.

Recent advancements in gasification technology have enhanced the viability of hydrogen production from biomass, with fluidized-bed gasification and supercritical water gasification techniques at the forefront due to their efficiency and reduced environmental impact. Research has shown that supercritical water gasification, in particular, facilitates the near-complete conversion of biomass to hydrogen, offering a cleaner alternative to traditional gasification methods [1, 2]. This review builds upon existing literature and employs a robust decision matrix to evaluate and compare these technologies, aiming to establish a strategic roadmap for their implementation within Colombia's burgeoning renewable energy sector.

2 Material and Methods

A systematic literature review was conducted to catalog existing gasification technologies, including fixed-bed, fluidized-bed, and circulating-bed systems. Selection criteria were established based on energy efficiency, synthetic gas quality, environmental impact, operational and maintenance costs, technological maturity, and safety indices. Subsequently, a comprehensive comparative analysis was performed using a decision matrix with a weighted scale, informed by a bibliometric review. This approach facilitated the development of a technological roadmap, aimed at guiding the selection and future implementation of the most suitable technology for sustainable hydrogen production, with a focus on achieving high safety standards and technological efficacy.

3 Results and Discussion

Following an exhaustive review of gasification methods for lignocellulosic biomass aimed at hydrogen production for energy purposes, various relevant technologies were identified, including different types of fixed-bed and fluidized-bed gasification, as well as plasma and supercritical water gasification (SCWG). A decision matrix was utilized to select the most appropriate technology, considering criteria such as technological maturity, gas composition, rate and efficiency of gas production, environmental impact, safety index, and investment cost.

Each technology was evaluated and scored on a scale of 1–5 for each criterion. Scores ranged from technological maturity (where 1 indicated a technology in the experimental stage and 5 a widely adopted commercial technology) to investment cost (where 1 represented a very high cost and 5 a very low cost). The results of the matrix revealed significant variations in the suitability of each technology for hydrogen production from lignocellulosic biomass, with some technologies showing higher efficiencies and lower environmental impacts, while others stood out for their operational safety or more attractive investment costs.

Four primary gasification technologies were assessed: fixed-bed gasification, fluidized-bed gasification, plasma gasification, and supercritical water gasification. The analysis of these diverse technologies for hydrogen production from lignocellulosic biomass highlights a range of energy efficiencies and operational characteristics. Fixed-bed gasification, including updraft and downdraft configurations, typically achieves 60–70% energy efficiency but produces low-quality gas with high tar content, requiring significant pretreatment and contaminant removal, costing between $50,000 and $105,000 for a 100-kWth system. Counter-current systems offer slightly better efficiencies (70–80%) and improved gas quality. Fluidized-bed systems, such as bubbling fluidized bed (BFB) and circulating fluidized bed (CFB), provide 65–75% and 70–75% efficiencies, respectively, with moderate gas quality and costs ranging from $80,000 to $140,000. Dual fluidized beds achieve 75–80% efficiency with high-quality gas. Plasma gasification and supercritical water gasification (SCWG) offer the highest efficiencies (80–85%), producing high-quality gas with low tar content; however, they involve higher capital and operational costs, estimated between $150,000 and $400,000 for SCWG and $200,000 and $400,000 for plasma systems. Each technology presents unique challenges in terms of emissions and waste management, with a general trend toward higher costs associated with higher efficiencies and cleaner gas production [3–6].

According to the total weighted scores, fluidized-bed gasification was identified as the most suitable technology for producing hydrogen from lignocellulosic biomass for energy purposes. It scored the highest total (4.15), indicating its strong performance across multiple evaluation criteria including technological maturity, gas composition, production rates, environmental impact, safety, and cost-effectiveness. This technology, known for its flexibility and energy efficiency, allows for better temperature control and increased surface area contact between the biomass and the gasifying agent, leading to higher biomass conversion into gas. Despite its higher cost and operational complexity compared to fixed-bed reactors, fluidized-bed gasification is favored for its ability to produce high-quality gas with high hydrogen content and low levels of contaminants.

In terms of emissions and safety, fluidized-bed gasification exhibits a superior ability to control and reduce pollutants such as nitrogen oxides (NOx), sulfur oxides (SOx), and particulates, which enhances its environmental suitability. This process allows for optimal air and biomass mixing, which not only improves combustion efficiency but also minimizes pollutants released into the environment. Furthermore, the technology's adaptability in handling various biomass moisture contents and its proven track record in agricultural and forestry residue conversion bolster its preference over the less mature fixed-bed gasification, which requires more stringent feed and temperature controls to achieve similar quality outputs.

Ultimately, considering the need for sustainable and efficient energy solutions in rural areas, fluidized-bed gasification stands out due to its operational advantages, higher pollutant control, and the robustness of its technology. It is suitable for large-scale hydrogen production projects, aligning with environmental goals and economic viability, whereas fixed-bed gasification may still be considered for smaller-scale applications.

4 Conclusion

The investigation into the technological roadmap for sustainable hydrogen production from lignocellulosic biomass via gasification has yielded a comprehensive assessment of multiple gasification technologies, highlighting the fluidized-bed gasification technique as the most advantageous in terms of energy efficiency, environmental impact, and economic feasibility. It is suggested that fluidized-bed gasification not only offers superior control over emissions but also provides a higher quality of syngas compared to other reviewed technologies, making it the most suitable option for future scaling and implementation in Colombia's context. Moreover, the economic analysis corroborates the potential for this technology to compete economically with traditional fossil fuel sources, given the right conditions, such as sufficient feedstock supply and supportive government policies in Colombia.

References

1. Ersöz A et al (2018) Int J Hydrog Energy 43(2):1081–1093. https://doi.org/10.1016/j.ijhydene.2017.11.017
2. Nanda S et al (2015) Int J Hydrog Energy 41(9):4907–4921. https://doi.org/10.1016/j.ijhydene.2015.10.060
3. Rodriguez Correa C, Kruse A (2018) J Supercrit Fluids 133:573–590. https://doi.org/10.1016/j.supflu.2017.09.019
4. Okonkwo EC et al (2021) Int J Hydrog Energy 46:35525–35549. https://doi.org/10.1016/j.ijhydene.2021.08.111
5. Kuo P-C et al (2021) Energy 223:120025. https://doi.org/10.1016/j.energy.2021.120025
6. Reddy SN et al (2014) Int J Hydrog Energy 39(13):6912–6926. https://doi.org/10.1016/j.ijhydene.2014.02.125

Testing the Biochemical Potential of Methane of Aqueous Glycerol Solutions Using Sludge from the Effluent Treatment System of Palm Oil Mills as Inoculum

Omar Alberto Ávila Rojas and Alexander Meneses-Jácome

Highlights

- The biological degradation of glycerol could be a significant source of biogas in the Colombian palm oil biodiesel sector for diversifying the national energy mix, closing energy and material cycles, and supporting future biorefining platforms.
- Sludge from the effluent treatment of palm oil mills (POME) is a promising source of inoculum for the anaerobic degradation processes of different agroindustrial effluents.

1 Introduction

The oil palm industry is a sector that could promote basic biorefinery schemas, dedicated to the transformation of biomass to biodiesel starting from a potential of 19.5% palm oil on a dry basis per ton of fresh fruit bunch (FBR) processed [1–4]. However, the design, operation, and control of the biological, chemical, and thermal processes used in a true biorefinery thought for this sector should imply a challenge due to their structural complexity and because some conversion routes have not yet a sufficient technological maturity for its implementation [1, 5–7]. These biorefineries arise from the need to minimize the amount of liquid wastes with high organic

O. A. Ávila Rojas (✉) · A. Meneses-Jácome
Circular Bioeconomy Research Center (ε-BiO): Biotechnology and Environment Group, Universidad Autónoma de Bucaramanga (UNAB), Bucaramanga, Colombia
e-mail: oavila845@unab.edu.co

A. Meneses-Jácome et al. (eds.), *The Circular Bioeconomy in Industry*, Waste as a Resource, https://doi.org/10.1007/978-3-031-84359-4_11

loads due to specialized molecules (e.g., glycerol, volatile fatty acids, etc.) being generated at different stages of the value chain, especially in the palm oil extraction and transesterification processes, which could provide energy to be recovered for the supply chain itself [1, 2, 8]. The generation of these effluents from oil extraction plants (POME, palm oil mill effluents) ranges between 0.5 and 1.2 m^3/ton of processed fruit bunch, which are continuously discharged causing severe water pollution [6, 9, 10].

Anaerobic digestion is a classical alternative for the treatment of solid and liquid waste in the agroindustry, which offers the possibility of adding value, particularly by obtaining bioenergy in the form of biogas rich in methane and even biohydrogen, as has been demonstrated with some new anaerobic-based technologies that eventually have to control this metabolic pathway [1, 2]. This study aims to establish a reliable technique to evaluate the true biochemical potential of methane (BPM) at the laboratory level, for the degradation of aqueous glycerol solutions as a model and relevant substrate of the palm oil and biodiesel sector using sludge from the treatment systems of POME, which are usually anaerobic and facultative ponds with high retention time and availability of quite active and mature sludge that seems to be an adequate source of inocula for glycerol degradation. Studying the controlled anaerobic degradation of glycerol is relevant because this substrate would assist and facilitate the degradation of more recalcitrant molecules present in POME and other effluents of the palm oil sector and it is necessary to understand the potential and limits of this capacity.

2 Material and Methods

Common procedures and methods were considered for all the experimentation.

2.1 Inoculum Preparation and Substrate

Anaerobic sludge extracted from a treatment system of POME located in the Santander region (Colombia) was used as inoculum. It was stored in a refrigerator at a temperature of 4 °C overnight and then was heated to 100 °C for 15 min. This inoculum was diluted with deionized water at a volume ratio of 60 and 90 mL, obtaining a concentration of volatile suspended solids close to 26,000 mL VSS/L. Analytical-grade glycerol was used as a substrate for methanogenic activity. Distilled water type 1 was used to prepare different solutions of this substrate. The glycerol was diluted with water at a volume ratio of 210 and 240 mL and a stoichiometric ratio of 1.217. Substrate-to-inoculum ratios of 2, 3, and 4 were used for the BPM activity tests.

2.2 *Analytical Methods and Experimental Design*

The inoculum to be studied was subjected to the following analyses: pH and temperature measurement using a multiparametric test; determination of the solids profile (total, total suspended, total dissolved, total volatile, volatile suspended, volatile dissolved) using standard methods (2540B, C, D, and E); and its chemical oxygen demand (COD) using the standard method 5220C. The evaluation of biogas generation potential and the measurement of the corresponding volume was carried out using the classical displacement method. This experiment was carried out in plastic vessels of 300 mL volume. These reactors contained a gas outlet hose submerged in a test tube inside a water bath at 23 °C, to displace the volume of water with the generation of biogas bubbles. The in vitro methane biochemical potential experiments (batch mode) were carried out using 3^2 experimental designs. In the case of glycerol, concentrations ranging from 8000 to 24,000 mg COD/L were used at standard environmental conditions (temperature of 23 °C and atmospheric pressure), for a methanogenic activity time of 10 days. Further to volume biogas generation (mL), the COD removal and pH were the measured variables.

3 Results and Discussion

Table 1 presents the characterization of the solids profile, total nitrogen, and COD measured in the investigation and compared to literature data. As evidenced in the table, all measured parameters are within the ranges under study.

Figure 1 shows the pH down from neutral to acid with a greater decrease when the inoculum is previously pretreated. Then, acidogenesis prevails on methanogenesis, and the inhibitory effect on biogas starts from the 5th to the 7th day. Figure 2 points out a higher biogas volume production as inoculum is pretreated, obtaining a maximum of 44 mLCH_4 at 10 days at 37 °C and 26 mL CH_4 at 23 °C. The highest biogas production is obtained at mesophilic conditions (37 °C) with pretreated inoculum as

Table 1 POME sludge characterization

Parameter	POME used in this study (mg/L)	POME—literature data (mg/L)	Method
Total solids	54,625	11,500–79,000	2540B—SM
Total suspended solids	41,911	5000–54,000	2540D—SM
Total volatile solids	34,816	9000–72,000	2540E—SM
Volatile suspended solids	25,978	7710–50,000	2540E—SM
COD	40,740	11,500–79,000	5220—SM
Total nitrogen	2600	180–14,000	Kjeldahl/titration

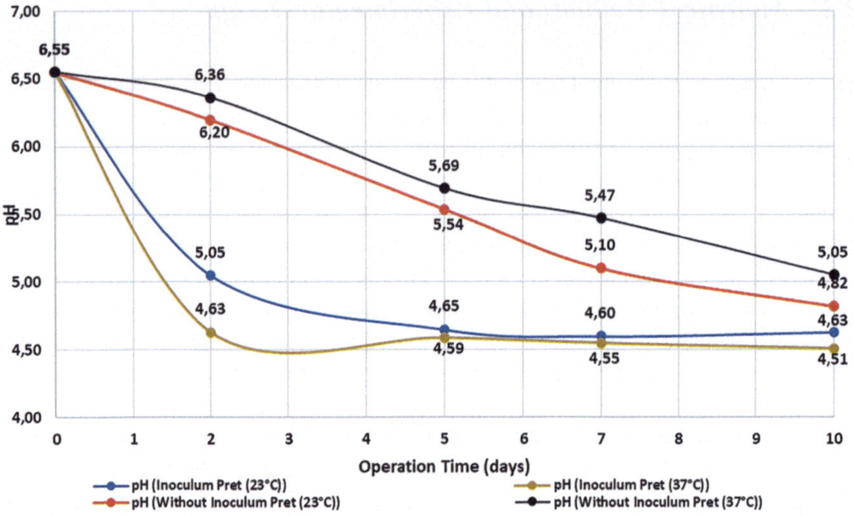

Fig. 1 pH trends vs inoculum pretreatment

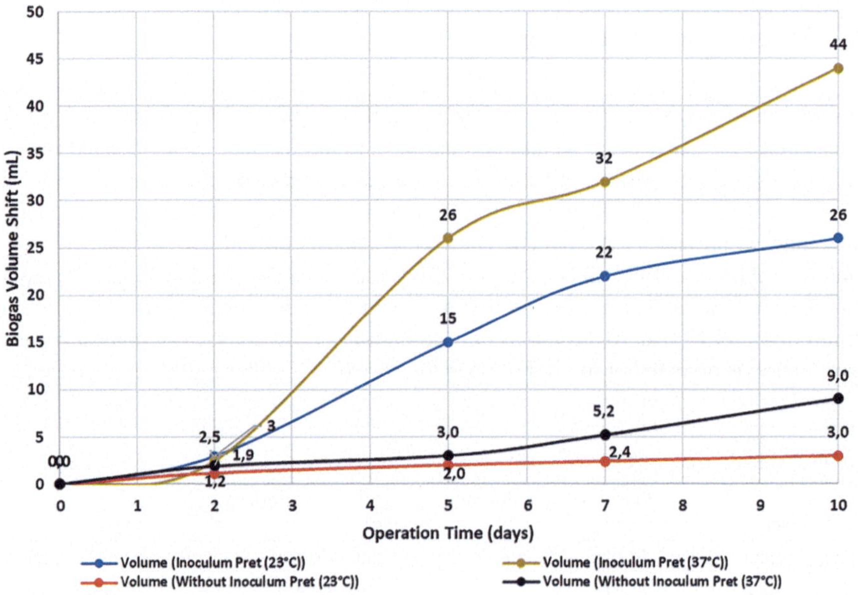

Fig. 2 Produced biogas vs inoculum pretreatment

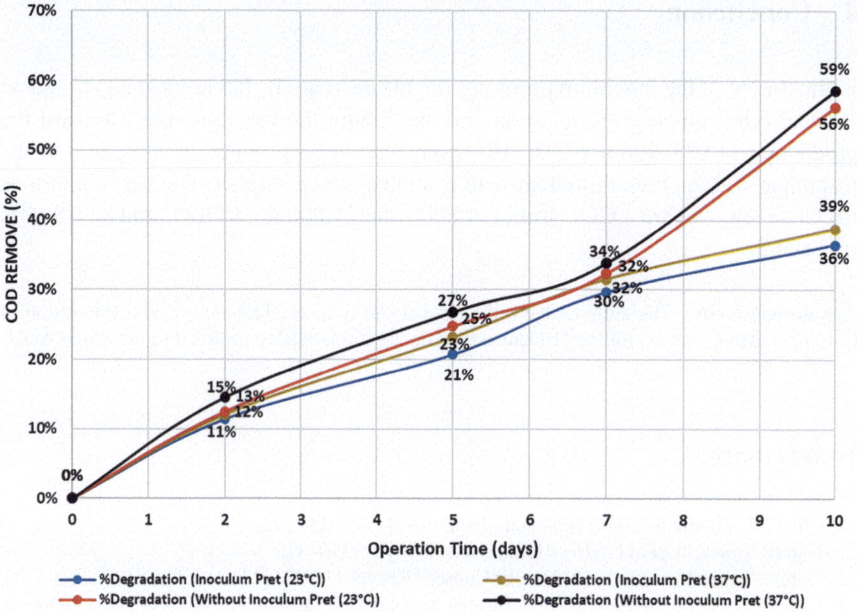

Fig. 3 COD removal (%)

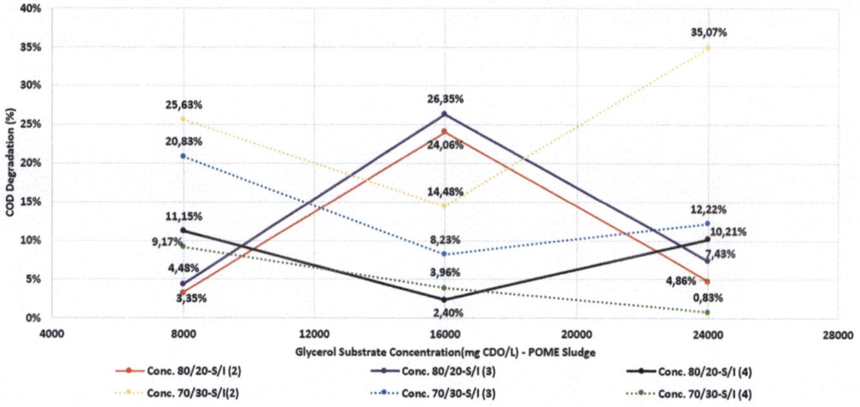

Fig. 4 COD removal (%) for different inoculum/substate ratios and substrate concentrations

observed elsewhere. Figure 3 reveals a higher COD removal when no inoculum pretreatment was performed, 59% at 23 °C and 56% at 37 °C for 10 days. Finally, Fig. 4 shows that the 70/30 concentration of the solution to inoculum/substrate ratio of 2 and 3 obtain higher COD removal at 8000 mg COD/L (20,83% and 25,63%) and 24000 mg COD/L (12,22% and 35,07%) of substrate. Concerning the concentration of 16000 mg COD/L of substrate, higher COD removal is obtained at 80/20 concentration and inoculum/substrate ratio of 2 and 3 (24,06% and 26,35%), respectively.

4 Conclusion

Pretreatment of the inoculum promotes the pH decrease by 0.5 units at 23 °C and 37 °C acidifying quickly the medium and stabilizing the reaction surely toward the production of CO_2 and not CH_4. However, the biogas production compared to the technique without inoculum treatment is almost 6 times higher. The best conditions for COD removal are a COD initial of 8000 and 24,000 mg COD/L and an I/S ratio of 2 and 3.

Acknowledgments The authors thank the Ministry of Science, Technology, and Innovation of Colombia (MinCiencias) for the "Bicentenario" doctoral scholarship awarded to Mr. Omar Ávila.

References

1. Chew TL, Bhatia S (2008) Bioresour Technol 99(17):7911–7922
2. Garcia-Nunez JA et al (2016) Biomass Bioenergy 95:310–329
3. Garcia-Nunez JA et al (2016) Resour Conserv Recycl 110:99–114
4. García, Varón D, Camilo J (2019) Turning an oil palm mill into a biorefinery: the passage of technologies through the valley of death. Palmas 40:76–103
5. Ali AM, Othman MR, Shirai Y, Hassan MA (2014) J Clean Prod 91:96–99
6. Peñuela DTR, Contreras NER, Garcia Nuñez JA (2014) Ceniavances 176:3–6
7. Quintero LA, Torres CI (2019) Ing USBMEd 10:8–18
8. Kamyab SH et al (2018) Palm oil
9. Reyes EI (2019) Estudio de la Generación y Composición de biogás en tratamiento de efluentes con biodigestores carpados en la industria de extracción de aceite de palma mediante simulación," Universidad Autonoma de Bucaramanga (in Spanish)
10. Malacatus-cobos PN, Llerena-carrera GM (2017) Cien Técn Aplic 3:459–469

Part II
Circular Bioeconomy Concepts and Bioeconomy for Food Security

Perspective on Coupled Agricultural Systems for a Profitable and Sustainable Production of High-Quality Food Proteins

Lukas Simon Kriem and Adna Saletovic

Highlights

- Global population growth, increasing affluence, urbanization, and evolving dietary preferences are driving diversification in protein production.
- As a fundamental building block of life, protein is crucial for a balanced diet, supplying essential amino acids for growth and tissue repair.
- A critical opinion on the challenges and opportunities in plant-based sources and emerging alternatives (including insects, fungi, plants, and microalgae) is presented.
- The potential for integrating by-products and waste streams across these different production systems, enhancing resource efficiency, sustainability, and profitability is discussed.

1 Introduction

Proteins are vital for life. Simultaneously, there is a deficiency of proteins in the worldwide food sources. Because of severe weather patterns due to climate change as well as contamination of soil and water from pesticides and fertilizers, this scarcity may rise in the upcoming years. The world population is projected to grow with a resulting increased demand for quality food proteins [1]. This, added to the obligation of improving sustainability and health, sets huge challenges to modern agriculture. One solution to address this issue is to introduce new protein sources that are

L. S. Kriem (✉) · A. Saletovic
Fraunhofer Institute for Interfacial Engineering and Biotechnology IGB, Stuttgart, Germany
e-mail: lukas.kriem@igb.fraunhofer.de

A. Meneses-Jácome et al. (eds.), *The Circular Bioeconomy in Industry*, Waste as a Resource, https://doi.org/10.1007/978-3-031-84359-4_12

sustainable and can serve as an alternative to animal-derived foods for the general population. Conventional farming practices, mainly based on maximization of yields and profitability, often result in environmental deterioration and reduction in nutritional value. With these urgent issues in mind, it can only be tackled with an approach at the systems level, integrating agricultural practices with the conservation of natural resources in an economically feasible, sustainable manner [2].

Sustainable agriculture is one approach that focuses on the restoration of ecosystem health while ensuring high productivity of protein alternatives [2]. Such a practice intends the interconnection of different agricultural structures, aiming to enhance crop production by incorporating regulatory and supportive ecosystem services. Professionals have been advising for a long time to transition to different protein sources, not just to reduce or completely replace animal-based food intake, but also to support sustainability efforts [3]. Utilizing alternative sources and maximizing their potential in the food industry necessitates the development of new agricultural systems for sustainable production, along with the effective utilization of by-products and waste material. Furthermore, a coupled agricultural system approach can provide in-depth solutions to the challenge of producing sustainable, high-quality proteins [4]. Two factors need to be considered in terms of coupled agricultural systems: (1) They need to be in line with sustainable cultivation approaches and (2) they need to produce profitable products. Hence, different conditions need to be evaluated for their potential to integrate different cultivation systems in future protein production. This opinion aims to summarize key considerations for the establishment of future protein systems with the focus on achieving profitability by integrating different protein production systems through energy and waste material coupling.

2 Selection of Protein-Production Systems

Alternative protein sources have the potential to reduce the overall demand for traditional protein sources and provide high-quality essential amino acids needed for human consumption. Besides, these sources of proteins have more desirable processing traits, which open them up to the food industry [5]. Here, four promising alternative protein systems have been considered as an example of coupling different agricultural systems. The proposed approach will comprise the production of various plant, insect, filamentous fungi, and microalgae protein-rich sources in controlled agricultural systems. This will also include the complete exploitation of all by-products from production to develop additional protein sources.

The choice of these protein sources is based on their potential to enhance sustainable and high-quality protein production. In fact, alternative protein sources such as microalgae, insects, fungi, and wheatgrass from vertical farming offer several advantages over traditional protein sources.

Microalgae is a highly productive producer of proteins, cultivable in photobioreactors all year round using less land and water than other conventional crops. Insects like mealworms, rich in complete proteins, can also be raised in contained systems.

Filamentous fungi could be grown in bioreactors to produce high-quality mycoprotein. Vertical farming of wheatgrass produces protein-rich greens with a fraction of the resources used for crops in the field [6].

Microalgae, exemplified by the primary photosynthetic organism *Chlorella vulgaris*, are recognized for their high efficiency in protein synthesis of up to 55% (w/w), while traditional agricultural crops lag behind in terms of productivity and resource efficiency. Grown under controlled photobioreactor conditions, microalgae can achieve continuous growth all year round, greatly reducing land and water use. This characteristic makes microalgal cultivation a promising alternative to conventional agriculture, especially in regions where arable land is limited or where environmental conditions impose a need for reduced usage of resources. Some species of algae also contain high levels of vital amino acids, further solidifying their position as a very important and complete protein provider [7].

Insects, in particular the species commonly known as mealworms (*Tenebrio molitor*), represent a possible and sustainable alternative source of protein, characterized by a high feed conversion efficiency. Mealworms produce complete proteins and can be farmed in contained, small systems that allow for control over environmental variables, minimizing the impact of climate variation on production outcomes. In addition, the farming of insects requires less resources—including feed and water—than traditional livestock farming, with its process also generating lower emissions of greenhouse gases. Additionally, previous works have come to the conclusion that protein content per used land is increased compared to conventional protein production [8]. This sustainable source of protein is particularly compatible with a circular economy framework, more so when using organic waste streams as feed inputs [9].

A few filamentous fungi, such as *Fusarium venenatum* or *Flammulina velutipes*, have been exploited for mycoprotein production using submerged fermentation in bioreactors. Growth of the filamentous fungi is described to be characterized by efficient resource utilization with high-protein biomass that has sensory similarities, texture, and nutritional richness analogous to meat, thus posing a very viable alternative to plant-based protein substitutes. This has the added advantage of mycoprotein production systems being scalable vertically, further optimizing land use, and a controllable fermentation process to enhance the quality and yield of protein [10].

In the final analysis, crops such as wheatgrass (*Triticum aestivum*) could be cultivated using vertical farming, showing it is feasible to produce nutrient-dense, protein-rich greens in limited areas with minimal environmental impacts. Vertical agricultural systems use artificial lighting, hydroponic methods, and controlled-environment regulation to create optimal conditions for growth, thereby significantly improving crop productivity per spatial unit compared to traditional, open-field systems. The result is an efficient way of producing protein-rich biomass, using only a fraction of the water and land resources of traditional crops. This characteristic makes this method particularly attractive for urban agriculture and sustainable food systems [11].

Enclosed growth systems, such as vertical farming for plants, insect farming for mealworms, bioreactors for fungi, and photobioreactors for algae, offer significant advantages over traditional agricultural methods. The main advantages include that these sources of protein can be produced year round, independently of climatic conditions, hence potentially leading to a highly efficient and resilient food production system [12]. Vertical farming, for example, uses only 5% of the traditional amount of water and cuts fertilizers by 50%, while completely eliminating the use of pesticides. A great reduction in the use of resources proves the sustainability and environmental benefits of these alternative ways of producing proteins [13].

Furthermore, the contained nature of these cultivation systems allows for very tight control over the growing environment, hence creating optimum conditions for the improvement and maturation of the selected protein sources. This enhances the overall productivity, yield, and quality of the proteins harvested to a great extent, thereby making it immensely attractive for integration into the food industry [14].

3 Coupling of Production Systems

Integrating these distinct protein cultivation methods within a coupled agricultural system establishes an optimized framework for resource management, waste reduction, and energy efficiency. By uniting different modes of protein production into a single, interconnected ecosystem, this approach fosters a "closed-loop" agricultural model, wherein by-products from one process serve as valuable inputs for another [15].

A crucial component in advancing sustainable protein production involves optimizing the utilization of waste and water by-products derived from energy production within the four cultivation systems. An example of the possible interconnection of the considered systems in terms of energy and resource coupling is shown in Figs. 1 and 2. By strategically incorporating these by-products, sustainable and cost-effective cycles can be established throughout the value chain, enhancing the ecological and economic viability of alternative protein production. In the context of vertical farming, not only are plant-based proteins produced efficiently, but all plant components are fully utilized, serving as substrates or growth media for secondary systems including insects as well as cultivation of fungi. Furthermore, waste heat generated from various processes can be repurposed, such as providing air conditioning within enclosed insect farming units, thereby enhancing energy efficiency [16].

Further, nutrient-rich effluents generated during algae cultivation can be recycled to enrich the nutrient profile of hydroponic solutions used in vertical farming systems for crops like wheatgrass [17]. Similarly, the reuse of nutrient solutions within a system such as microalgae production or production of filamentous fungi reduces fertilizer demand. Simultaneously, the carbon dioxide released during fungal fermentation processes can be channeled to microalgal photobioreactors, providing essential carbon sources that promote algal growth [18]. This circular approach not only minimizes resource input and waste output but also promotes a self-sustaining

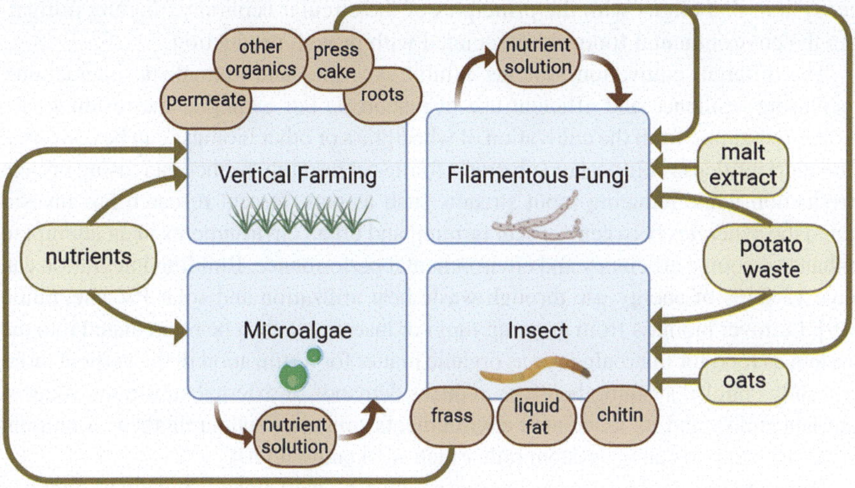

Fig. 1 Schematic of coupling potential of different waste streams and integration in another system. Red arrows indicate the reuse within a system. Brown arrows indicate reuse across all considered protein-production systems. Red elements indicate a waste product. Brown elements indicate an external resource

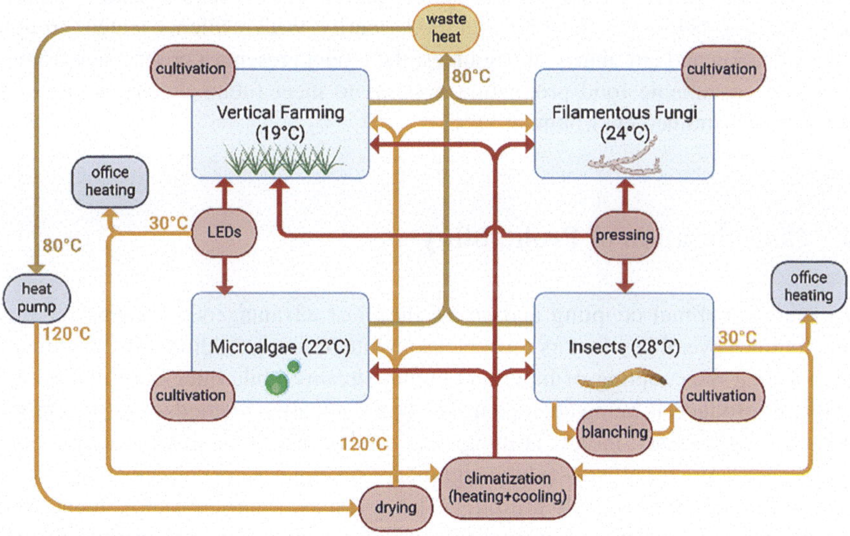

Fig. 2 Schematic of coupling potential of different energy streams and integration in another system. Red arrows indicate the electrical energy input into a system. Yellow arrows indicate the thermal energy input into a system with specific heat requirements for different segments. Brown arrows indicate waste heat from processing products. Red elements indicate energy requirements. Brown elements waste heat. Gray elements indicate external systems not required for the functionality of an individual system but required for the best coupling potential

ecosystem that aligns with the principles of the circular economy, thereby mitigating the environmental footprint associated with protein production.

The different cultivation systems exhibit synergies and mutualistic interactions, increasing resilience and efficient use of resources. For example, the residual substrates remaining from the cultivation of wheatgrass or other biomass can be converted into nutrient-dense compost or substrates for insect farming, further increasing protein production while reducing input streams such as oats. Recent research has investigated the synergies between vertical farming and urban environments in an attempt to enhance resource efficiency and environmental performance. Building integration can save 12–51% of energy use through waste heat utilization and solar PV integration [19]. Leftover biomass from growing fungi or insects can then be recirculated into the system as food for microalgae or as organic matter for fertilization in the vertical farm. A tightly coupled system, therefore, reduces demands of external substrates, reduces nutrient runoff and its associated environmental impacts, and strengthens a nutrient cycle that supports an ecosystem rather than weakening it [20].

Diversity in protein sources can give a coherent, well-linked framework that can raise resilience, adaptability, and general dependence in food production [21]. Diversification of the sources of this system makes risks associated with environmental change, market demand shifts, and possible supply chain disruptions less likely, thus ensuring stable and sustainable supplies of high-quality protein. Such an integrated model can enable adaptive responses to changes in resource availability or exogenous pressures, making the system more robust to unexpected perturbations and thus helping to ensure global food security. With resilience added through diversification and circularity of resources, the model presents a progressive framework for sustainable food production systems to meet future dietary needs in a resource-constrained environment [22].

4 Consideration of Profitability

To achieve optimal coupling and create the most advantageous systems for this integration, several key factors must be taken into account. Additionally, the aim of maximizing the coupling of input and output streams while integrating the profitability of products is being attempted. The four cultivation methods—vertical farming for wheatgrass, insect farming for mealworms, fungal bioreactors, and photobioreactors for algae—exhibit significant synergies and complementary interactions that can be effectively utilized [23]. For instance, nutrient-rich effluent from the production of algae can be utilized as fertilizer in indoor vertical farms of wheatgrass cultivation. This would create a closed-loop system whereby, through some means, the by-products of one system result in inputs to another.

In this way, resources would be utilized most efficiently while waste generated would be minimal, thus improving the revenue of products. Additionally, carbon dioxide emitted from fungal bioreactors can be captured and supplied to the microalgae, adding more potential toward making the system even more sustainable [24].

Another idea could be based on the utilization of alternative sources of proteins and their derivatives in food. For instance, high-quality protein (and also other by-products in the downstream processing) from wheatgrass can become a great constituent in many food products while proteins from insects can be used in animal feeds or pet foods [25]. The fungi proteins also have unique flavor and texture profiles, possibly applicable in meat substitutes or specialty food items. Algae-derived proteins are nutrient-dense and would find applications in functional food.

Integration of all products and by-products from the four cultivation systems can optimize the overall profitability and economic sustainability of the coupled agricultural system [26]. The nutrient-dense, plant materials and waste streams produced from the distinct systems can serve as inputs to supplementary, value-added processes, including the production of biofuels or biochemicals, or possibly even as soil amendments for organic farming practices. Specifically, the holistic approach toward resource use and waste management can help in developing a circular and sustainable food production system [27]. For instance, the wheatgrass grown in the vertical farming system can be processed to extract high-value protein concentrate, potentially sold as a premium ingredient to food companies. The remaining wheatgrass after initial processing, which is the residual mass left after protein extracts have been separated, holds significant value as a nutrient source [28]. Rather than discarding this mass, it can be transformed into fertilizer for an insect farming system with mealworms, serving as nutrient-rich broken feed. Furthermore, another consideration is the use of its byproduct, wheatgrass biomass, to become a substrate for the insect farming operation to potentially become an inexpensive feed for mealworms.

The integrated, closed-loop microalgae cultivation and vertical farming system can offer recycling of nutrients and reductions in wastes for increased sustainability and resource efficiency. Microalgae can provide natural fertilizers during vertical farming, such as nutrient-rich effluent produced by algae during cultivation being used to fertilize wheatgrass [29]. This mutually dependent use of resources minimizes the need for chemical fertilizers and reduces input costs as well as environmental stress.

It also encourages circularity as it turns the waste outputs of one subsystem into energy inputs in another. This interconnected system facilitates resource optimization, as shown by the ability to capture carbon dioxide emissions from fungal bioreactors and redirect them to stimulate the growth of algae [30]. Algae cultivation not only assists in effective carbon dioxide recovery by utilizing its gaseous emissions as feedstock but also produces biomass that can be used for various applications, such as biofuels, food additives, and animal feeds.

The growing incorporation of mealworm protein in pet foods has introduced a new green protein source for an innovative approach. Similarly, fats derived from insects, particularly those from mealworms, extend much broader applications beyond food production [31]. They may be converted into certain greases for useful utilization in several industries, including lubricants for machinery in the bio-based manufacturing approach [32]. By incorporating other revenue streams of by-products, the integrated production system becomes significantly more robust and financially sustainable as the sale of edible insect-based fats creates additional revenue streams. Additionally, the nutrient-rich effluents generated by an algae cultivation system in the same facility can be redirected to nourish the wheatgrass in the vertical farm [33]. This nutrient recycling not only increases the whole system's sustainability by lowering the dependence on external fertilizers, but it also permits efficiency inside a closed-loop environment.

Lastly, bioreactors that utilize fungi will further enhance their worth by producing biochemical substances that have wide applications in the pharmaceutical and chemical industries. Conventional fungal culture has led to the discovery of compounds that have antioxidant, antimicrobial, and therapeutic properties, which are often used in various pharmaceutical formulations [34]. In addition to pharmaceutical applications, fungal metabolites from natural sources can also find other applications in chemical industries. This diversification of outputs in the integrated biotechnology system by including fungi-based products will further strengthen the economic resilience of the facility and support biotechnological innovation across different industries.

This closed-loop integration of agriculture and ecology demonstrates how to maximize the benefits from the land while simultaneously enhancing its resilience. It provides an illustration of how circular economy flows of waste-to-resource are possible within controlled-environment agriculture, which can generate economic and environmental benefits. Such integrated systems can reduce reliance on inputs, conserve water and nutrients, and may help to provide pathways toward greater regionalized sustainable food systems.

In summary, the integration of different alternative protein sources introduces an integrated circular bioprocessing system that showcases a synergistic mechanism where the output of one unit serves as the input for the next upstream unit. These circular processes emphasize more efficient resource utilization, reduced waste generation, and the production of a wide variety of high-value products. Although maximum coupling of streams may serve as one promising approach toward the profitability of systems, other factors should be considered. This could involve the consideration of only some systems to reduce energy demands or to have an optimal loop of input and output streams. Additionally, capital costs as well as operational costs, specifically energy costs can vary significantly depending on the location of such production systems. In terms of combining sustainability and profitability, the best possible coupling of individual processes will be required. This opinion is based on the concept of coupling individual systems across their boundaries not considering further profitability options such as capital costs or geographic implementation.

5 A Way Forward

Economic viability analysis is essential for the described industrial protein production, combining numerous financial metrics and issues of renewability or sustainability. Financial metrics such as net present value (NPV), return on investment

(ROI), or internal rate of return (IRR) play a key role in profitability calculation and understanding the economic feasibility of industrial protein production [35]. Cost–volume–profit (CVP) and break-even analysis, both of which are considered supplementary analyses, determine production levels required to generate a profit. Looking beyond financial indicators, sustainability profitability should integrate environmental and social consequences, leveraging life-cycle assessment (LCA) and social return on investment (SROI) to weigh economic objectives against sustainability imperatives. Efficiency (i.e., low raw materials costs) presents a key factor, as well as supply chain and scalability [36]. Predictive analytics and process simulation, also related to the simulation of production facilities, are examples of digital tools that can help accurately forecast costs and pinpoint the optimal pathway for production under different conditions.

Coupling processes in protein production refer to the integration of different stages (fermentation, separation, purification, and packaging) with the aim to reduce costs and increase sustainability [37]. Waste from one system as an output stream can act as an input for another system through process integration, which improves resource use and reduces environmental impacts. Technological compatibility leads to standardization and regulatory compliance for safety and quality key drivers of successful coupling. Elements such as process simulation software and control systems are essential for ensuring tight integration and for identifying and addressing potential issues that may occur during the scaling-up process.

Choosing the right geographic location directly plays a critical role in your profit margin for variable factors such as labor, energy rates, and regulatory environment. On-site access to renewable energy can lower operating costs while improving both economy and sustainability [38]. Moreover, local production close to markets or distribution networks reduces transportation costs and increases general efficiency [39]. Therefore, its location is deliberately chosen to create synergies where the logistics can be performed in the most effective manner providing savings and achieving financial as well as sustainability targets.

As technology advances and becomes more established, it is essential to shift toward prioritizing the affordability of protein to meet the needs of not just high-income consumers, but also to ensure global food security by making nutritious food more accessible. Meeting the increased global demand for food requires innovative strategies such as utilizing coproducts, coupling different protein-production systems, reducing waste, and adopting new protein technologies alongside existing industry practices. This necessitates collaboration across all industries to develop a more comprehensive solution. A sustainable food system is more attainable when we collaborate instead of working individually moving beyond system boundaries. This requires collaboration across cultivation systems to become a competitor to conventional protein sources in the future.

Funding The work presented in this chapter stems from the research conducted for the Fraunhofer Flagship Project "FutureProteins," funded by the Fraunhofer-Gesellschaft.

References

1. Smetana S et al (2023) Resour Conserv Recycl 190:106831. https://doi.org/10.1016/j.resconrec.2022.106831
2. LaCanne CE, Lundgren JG (2018) PeerJ 6:e4428. https://doi.org/10.7717/peerj.4428
3. Azizi R et al (2024) Crit Rev Food Sci Nutr. https://doi.org/10.1080/10408398.2024.2365339
4. Roberts DP, Mattoo AK (2019) Front Sustainable Food Syst 3:72. https://doi.org/10.3389/fsufs.2019.00072
5. Ismail BP et al (2020) Anim Front 10(4):53–63. https://doi.org/10.1093/af/vfaa040
6. Schweiggert-Weisz U et al (2020) Curr Opin Food Sci 32:156–162. https://doi.org/10.1016/j.cofs.2020.08.003
7. Chen C-Y et al (2015) Biotechnol J 10(6):905–914. https://doi.org/10.1002/biot.201400594
8. Hasnan FFB et al (2023) Food Secur 12(23):4243. https://doi.org/10.3390/foods12234243
9. Spang B, Francis K (2013) Insects as food: assessing the food conversion efficiency of the mealworm (Tenebrio molitor). https://www.semanticscholar.org/paper/Insects-as-food-%3A-Assessing-the-food-conversion-of-Spang-Francis/73635b3b284e74236141dee6d5fbb22899a4b225. Accessed 18 November 2024
10. Reihani SFS, Khosravi-Darani K (2018) Appl Food Biotechnol 5(4):4. https://doi.org/10.22037/afb.v5i4.23139
11. Asseng S et al (2020) Wheat yield potential in controlled-environment vertical farms. Proc Natl Acad Sci 117(32):19131–19135. https://doi.org/10.1073/pnas.2002655117
12. Torres-Tiji Y, Fields FJ, Mayfield SP (2020) Biotechnol Adv 41:107536. https://doi.org/10.1016/j.biotechadv.2020.107536
13. Despommier D (2010) The vertical farm: feeding the world in the 21st century. https://www.semanticscholar.org/paper/The-Vertical-Farm%3A-Feeding-the-World-in-the-21st-Despommier/babed74420c8f706f218014570c6b12c6e020d1d. Accessed 18 November 2024
14. Daniels A et al (2023) IFAC-PapersOnLine 56(2):9887–9893. https://doi.org/10.1016/j.ifacol.2023.10.666
15. Durkin A et al (2022) Curr Res Biotechnol 4:365–376. https://doi.org/10.1016/j.crbiot.2022.09.001
16. Rasool K et al (2023) Rev Environ Sci Biotechnol 22(2):527–562. https://doi.org/10.1007/s11157-023-09651-6
17. Khan S et al (2023) Curr Opin Green Sustainable Chem 41:100797. https://doi.org/10.1016/j.cogsc.2023.100797
18. Dsouza A et al (2021) Sustainability 13(5):2471. https://doi.org/10.3390/su13052471
19. Blom T, Jenkins A, van den Dobbelsteen A (2023) Front Sustainable Food Syst 7:1227672. https://doi.org/10.3389/fsufs.2023.1227672
20. Martin M, Weidner T, Gullström C (2022) Front Sustainable Food Syst 6:849304. https://doi.org/10.3389/fsufs.2022.849304
21. MacFall J et al (2015) J Environ Stud Sci 5(4):608–622. https://doi.org/10.1007/s13412-015-0321-1
22. Weindl I et al (2020) Glob Food Sec 25:100367. https://doi.org/10.1016/j.gfs.2020.100367
23. Kannan H, Mesmer BL, Bloebaum CL (2017) Syst Eng 20(1):21–44. https://doi.org/10.1002/sys.21377
24. Kumar A et al (2010) Trends Biotechnol 28(7):371–380. https://doi.org/10.1016/j.tibtech.2010.04.004
25. De Souza-Vilela J, Andrew NR, Ruhnke I (2019) Anim Prod Sci 59(11):2029. https://doi.org/10.1071/AN19255
26. Shang YC, Costa-Pierce BA (1983) J World Maricult Soc 14(1-4):523–530. https://doi.org/10.1111/j.1749-7345.1983.tb00104.x
27. Jurgilevich A et al (2016) Sustainability 8(1):69. https://doi.org/10.3390/su8010069
28. Rana S et al (2021) Microorganisms 9(5):1004. https://doi.org/10.3390/microorganisms9051004

29. Meles Y et al (2020) Sci Total Environ 737:139631. https://doi.org/10.1016/j.scitotenv.2020.139631
30. Sørensen IM, Tarp SH (2023) Sustain Sci 18(1):203–222. https://doi.org/10.1007/s11625-023-01242-x
31. Lorette B and Sanchez L (2022) New lipid sources in the insect industry regulatory aspects and applications OCL 2922. https://doi.org/10.1051/ocl/2022017
32. Kee PE et al (2023) Insect biorefinery: A circular economy concept for biowaste conversion to value-added products Environmental Research 221115284. https://doi.org/10.1016/j.envres.2023.115284
33. Cowan N et al (2022) Integration of algae treatment with hydroponic crop waste to reduce impact of nutrient waste streams Abstract Journal of Sustainable Agriculture and Environment 1(3) 203–215. https://doi.org/10.1002/sae2.v1.3, https://doi.org/10.1002/sae2.12025
34. Olicón-Hernández DR et al (2017) Overview on the Biochemical Potential of Filamentous Fungi to Degrade Pharmaceutical Compounds Frontiers in Microbiology 8. https://doi.org/10.3389/fmicb.2017.01792
35. McNulty MJ et al (2022) Recombinant Proteins in Plants Methods and Protocols Technoeconomic Modeling and Simulation for Plant-Based Manufacturing of Recombinant Proteins Springer US New York NY 159–189
36. Fletcher AJ et al (2024) Modeling the feasibility of fermentation-produced protein at a globally relevant scale Frontiers in Sustainable Food Systems 8. https://doi.org/10.3389/fsufs.2024.1419259
37. Kiss AA et al (2014) A systems engineering perspective on process integration in industrial biotechnology. https://doi.org/10.1002/jctb.4584
38. Day CJ (2022) Why industrial location matters in a low-carbon economy Structural Change and Economic Dynamics 63:283–292. https://doi.org/10.1016/j.strueco.2022.10.004
39. Tajbakhsh A and Shamsi A (2019) A facility location problem for sustainability-conscious power generation decision makers. Journal of Environmental Management 230:319–334 https://doi.org/10.1016/j.jenvman.2018.09.066

Bioeconomy: Agro and Food Security

Maria del Pilar Márquez Cardona

Highlights

- Food security is a global challenge that largely falls on the agricultural sector. Armed conflicts, high fertilizer prices, and the increased consumption of ultra-processed foods are some examples that jeopardize food security.
- Bioeconomy repositions agriculture as a development axis, adding value to products and residues generated in agriculture and bringing opportunities for innovation and job creation.
- In Latin America, the transition toward a sustainable circular bioeconomy includes strengthening capacities in science, technology, and innovation, promoting technological cooperation, and implementing robust, effective, and long-term public policies and strategies, integrating institutions from the public sector, private sector, and academia.

1 Introduction: The Challenges of Food Security

Food security, understood as the sufficient and stable availability of food, its safety, timely access, and biological utilization for all people to meet their nutritional needs and preferences to achieve a healthy and active life (FAO), continues to be one of the great challenges for humanity. According to FAO [1], around 735 million people worldwide suffer from food insecurity, despite the constant increase in crop yields.

Armed conflicts are one of the main triggers of hunger, as they directly impact the supply of food and water, especially in the most vulnerable populations,

M. del Pilar Márquez Cardona (✉)
Pontificia Universidad Javeriana, Bogotá, Colombia
e-mail: marquez.maria@javeriana.edu.co

A. Meneses-Jácome et al. (eds.), *The Circular Bioeconomy in Industry*, Waste as a Resource, https://doi.org/10.1007/978-3-031-84359-4_13

children, and women [2]. Moreover, 70% of the people suffering from hunger in the world live in areas where there are wars and violence.

Fertilizer prices have been shown to have a significant impact on food security. High input prices influence food availability as they lead to a drop in food production, such as corn, soybeans, rice, and wheat, which constitute the main caloric intake for humanity [3].

Another manifestation of food insecurity is malnutrition. In many developing countries, malnutrition remains a significant concern, with an estimated 390 million adults and 190 million children and adolescents being underweight in 2022. On the other hand, in both developed and developing countries, the increase in the consumption of processed foods rich in fats and sugars, along with sedentary lifestyles, has contributed to the rise in overweight and obesity, even among low-income populations [4].

All these challenges largely fall on the agricultural sector. Soils, water, and plant genetic resources are key inputs for sustainable food production. Factors such as population growth, which in turn intensifies urbanization and deforestation, have brought significant pressures on agri-food systems. It is essential to increase crop yields without expanding the agricultural frontier; likewise, the restoration of degraded lands, proper management of water resources, and the development of new varieties resistant to biotic and abiotic factors are imperative.

2 Bioeconomy: Some Approaches and Contexts

2.1 Bioeconomy and the Agricultural Sector

The intersection between circular bioeconomy and food security emerges as a key strategy to address contemporary challenges in food production. In this sense, bioeconomy repositions agriculture as a development axis, adding value to products and residues generated in agriculture and bringing opportunities for innovation and job creation. Bioeconomy, as a productive and economic strategy, promotes the conservation and efficient, sustainable use of biological resources and reduces our dependence on fossil fuels [5].

To achieve sustainable agricultural production, it is necessary to manage biodiversity and ecosystem functions. Biodiversity plays a central role as a regulator of agroecosystem functions to meet a range of needs for both farmers and consumers [1]. Thus, bioeconomy, in addition to offering cost-effective solutions, must be inspired by and supported by nature, simultaneously providing environmental and social benefits and increasing the resilience of agroecosystems. To develop these solutions, it is necessary to understand the system and identify its functions and services (nature-based solutions), which involve science, technology, and innovation.

2.2 Bioeconomy in Latin America

Latin America and the Caribbean have abundant natural resources. According to data from CEPAL [6], this region holds over 50% of the world's agricultural potential and estimates that by 2050, more than 300 million hectares can be produced without affecting natural forests. Additionally, it is the center of origin and diversity of species that sustain food worldwide, and it hosts a multitude of species with pharmaceutical, nutraceutical, and cosmeceutical potential and harbors over 30% of the planet's freshwater. These circumstances place Latin America and the Caribbean in a position to benefit from a bioeconomy that can contribute to food security and the supply of renewable energy [7].

However, in Latin America and the Caribbean, most bioeconomy initiatives and strategies focus on biomass production, with little value added to the products. This production centers on crops such as sugarcane, coffee, corn, soybeans, palm oil, and fruits, and in most diagnoses of these productive chains, poor technological performance is identified as the most significant limitation to improving resource use efficiency [7].

Trigo et al. [8] described the opportunities in Latin America and the Caribbean that form the basis for the discussion that must take place in the region to close institutional, policy, and knowledge generation gaps, favoring the effective development of the bioeconomy. *Biodiversity*: In Latin America, there is a great diversity of underutilized crops that can be used as raw materials in phytotherapy, cosmetics, and food. Eco-intensification promotes conservation and precision agriculture by integrating clean technologies and biological processes. *Biorefineries and Bioproducts*: The region has growth potential in transforming raw materials into different product chains and reducing the cost of primary products. Biorefineries play an important role as they seek to improve the life cycle performance of products and enhance the environmental performance of these industries. Biotechnology offers innovative solutions to the agricultural sector and natural resource management, reducing environmental impacts. Within a bioeconomic approach, it is necessary to identify and manage ecosystem services since bioeconomy is a response to a long period of excessive resource use and an attempt to readjust behaviors in light of global challenges. In this sense, the integrity of the natural environment and the value of different flows in the system must be recognized. Last but not least, achieving greater efficiency in value chains is important, focusing on reducing postharvest losses and developing markets for innovative biological products.

All these opportunities highlight the need in Latin America and the Caribbean to strengthen capacities in science, technology, and innovation, promote technological cooperation, and implement robust, effective, and long-term public policies and strategies, integrating institutions from the public sector, private sector, and academia, to transition toward a sustainable circular bioeconomy [9].

2.3 *Bioeconomy and Family Farming*

In Latin America and the Caribbean, family farming is of great importance for food security, representing 80% of agricultural holdings and being the main source of agricultural and rural employment. Most of the food for internal consumption in the region is produced under this system [10]. However, this sector has one of the lowest productivity, less access to technologies and markets, and little capacity to mitigate the effects of climate change.

Bioeconomy represents an opportunity for family farming, increasing productivity through access to high-quality seeds, reducing postharvest losses, transforming biomass into small-scale bioindustrial products, reducing environmental impacts in production systems, and making rural employment more attractive to new generations [11]. It is necessary to develop policies that promote biotechnological opportunities to add value to biodiversity and biomass, which will result in reducing the vulnerability of family farming to climate change and improving the profitability of production units.

3 Conclusion

The integration of bioeconomy principles into the food production system is a promising pathway to enhance food security in the face of global challenges. By leveraging biotechnological innovations, promoting sustainable resource management, conserving biodiversity, and fostering economic opportunities, bioeconomy contributes to a more resilient, efficient, and inclusive food system. Policymakers, researchers, and stakeholders must collaboratively work toward implementing and scaling up bioeconomic initiatives to ensure a sustainable and secure future for global food production and consumption.

References

1. FAO (2023) El estado de la seguridad alimentaria y la nutrición en el mundo. https://www.fao.org/3/cc3017es/cc3017es.pdf (in Spanish)
2. Unicef. https://www.unicef.es/blog/desnutricion/por-que-sigue-habiendo-hambre-en-el-mundo. Consulted in March 2024
3. WFP: https://es.wfp.org/crisis-alimentaria-mundial. Consulted in March 2024
4. WHO: https://www.who.int/news-room/fact-sheets/detail/malnutrition. Consulted in March 2024
5. Global Bioeconomy Summit. https://gbs2020.net/wp-content/uploads/2021/10/GBS_2018_Report_web.pdf. Consulted in March 2024
6. Cepal (2011) Análisis comparativo de patentes en la cadena de producción de biocombustibles entre América Latina y el resto del mundo. En Diálogo de políticas sobre desarrollo institucio-

nal e innovación en biocombustibles en América Latina y el Caribe. Santiago de Chile, Chile: Comisión Económica para América Latina. (in Spanish)
7. Hodson de Jaramillo E (2014) Hacia una bioeconomía en América Latina y el Caribe en asociación con Europa. Editorial Pontificia Universidad Javeriana, p 149 (in Spanish)
8. Trigo EJ, Henry G, Sanders J, Shurr U, Ingelbrecht I, Revel C, Santana C, Rocha P (2013) Towards bioeconomy development in Latin America and the Caribbean. Bioeconomy Working Paper. ALCUE-KBBE
9. Misión de Sabios Colombia (2019) Colombia hacia una sociedad del conocimiento. Reflexiones y propuestas (in Spanish)
10. FAO (2014) Agricultura Familiar en América Latina y el Caribe: Recomendaciones de polìtica. Organización de las Naciones Unidas para la Alimentación y la Agricultura (in Spanish)
11. Fontagro (2016) Memoria del Seminario de Bioeconomía. https://www.fontagro.org/new/uploads/adjuntos/MEMORIA_BIOECONOMIA_FINAL_ESP.pdf (in Spanish)

BioInsectonomy: A Circular Agrifood Economy Based on Insects as Animal Feed

Karol B. Barragán-Fonseca

Highlights

- This study advocates for a circular economy involving insects to efficiently convert low-cost organic waste into high-quality animal feed.
- The project explores the potential of insects fed on organic waste for use as animal feed, offering valuable insights for developing sustainable practices in the global bioeconomy.

1 Introduction

As a result of ecological crises and economic pressure, the global agrifood system is currently pressed to produce food using more sustainable resources. In turn, increased demand for food due to population growth is increasingly affecting natural resources, as well as ecological systems, and leading to rising costs of agricultural inputs. Additionally, approximately 30–40% of food produced is lost along the production chain [1], resulting in high levels of organic waste. Therefore, there is a need to promote more sustainable food sources and efficient forms of production to respond to these challenges and contribute to meeting the sustainable development goals (SDGs) formulated by the United Nations Food and Agriculture Organization.

One such option involves transitioning towards circular food systems through practices and technologies that minimise the use of finite resources, involve the use of regenerative resources, prevent the leakage of natural resources from the food

K. B. Barragán-Fonseca (✉)
Faculty of Veterinary Medicine and Animal Sciences, Universidad Nacional de Colombia, Bogota Campus, Bogota, Colombia
e-mail: kbbarraganf@unal.edu.co

A. Meneses-Jácome et al. (eds.), *The Circular Bioeconomy in Industry*, Waste as a Resource, https://doi.org/10.1007/978-3-031-84359-4_14

system, and favour the recycling of resources, which furthermore adds value to the food system [2]. Additionally, alternative nutrient sources are required to reduce negative impacts on ecosystems and preserve biodiversity. In the pursuit of circular food systems, further exploring efficient, sustainable use of biological resources is essential. Of all types of organisms, insects in particular provide numerous ecosystem services and have high food conversion rates, low levels of greenhouse gas production, and low production costs [3, 4].

Several insect species may be fed on organic waste and, in turn, be used as biomass for animal feed, resulting in efficient resource use and low cost of inputs and production [5]. Furthermore, their waste—or frass—may be used as a biofertiliser and has been shown to enhance plant growth and health [6] while also contributing to the economies of small farmers and communities [7]. Finally, the use of insects in circular food systems contributes to several SDGs [8] (Fig. 1).

To maximise the efficiency of insects as recyclers in food systems, they should be fed organic waste or low-opportunity-cost feed (LOCF), consisting of food waste and crop residues and to a lesser extent by-products of the food industry [9]. Producing insects as food and bioconverters involves seeking resource efficiency in

Fig. 1 Benefits and impacts of insect production in circular food systems within the bioeconomy

animal feeding and breeding by identifying LOCF that maximise bioconversion efficiency, as well as insect species with high potential for bioconverting them; evaluating and optimising quality and safety of biomass and frass for use as animal feed and biofertiliser; and establishing safety practices for growth of this emerging area of animal production, which may significantly contribute to the bioeconomy.

One major concern is variability in the efficiency of insect bioconversion due to substrate variability [5]. Factors such as the substrate's nutritional composition strongly impact insects' development and body composition, frass composition and quality, the environmental footprint, and product safety [10]. Understanding insects' influence on nutrient flows and loss during bioconversion of LOCF is crucial for ensuring sustainable insect farming, which contributes to circular food systems [7]. This project investigates those organic by-products and insect species suitable for aquaculture. Additionally, we explore risk assessment strategies for sustainable insect meal to ensure that insect meal can and will be incorporated into fish feed formulations, helping to mitigate future protein shortages. Analytical tools for tracing insect feed will allow insect farmers to test insect-based products for compliance with EU legislation, which is currently the most developed in the world with respect to insects for food and feed.

2 Material and Methods

The project consists of three phases. We have completed the first phase and are currently finalising the second, with the third phase pending. *Phase 1. Selection of agricultural residues and insect species*: We designed a framework for evaluating 23 types of agricultural waste in Colombia, scoring them based on production volume, supply, bromatological and microbiological quality, toxicological aspects, potential for insect feeding, cost per kilogram, and other criteria. We also evaluated 11 insect species, scoring them based on space requirement, life cycle duration, uses in aquaculture, existing related legislation, ease of handling, reproductive performance, biodegradation capacity, resistance to environmental changes, and other criteria. *Phase 2. Insect performance and safety of meals using organic waste*: We formulated four diets based on different types of organic waste and assessed insects' productive performance and nutritional composition, as well as the safety of the resulting insect meal. *Phase 3. Fish performance*: We evaluated the productive performance of tropical cachama fish (*Piaractus brachypomus*) fed with insects in turn fed on organic waste.

3 Results and Discussion

Phase 1: Based on their scores mentioned above, we selected five types of LOCF for inclusion in insects' diets (mango, cassava, and carrot residues; pig manure, and coffee husks) and two agricultural by-products (guava meal and wheat bran). Three experimental insect diets were formulated by blending these ingredients. We selected two insect species for inclusion in the experiment: black soldier fly (*Hermetia illucens*) and mealworm (*Tenebrio molitor*). *Phase 2*: The greatest biomass of larvae was obtained using high levels of cassava residues and pig manure. Microbiological and toxicological analyses of the larvae confirmed the safety of insect meal derived from LOCF for inclusion in animal diets. However, it is essential to implement good practices and HACCP (Hazard Analysis and Critical Control Points) methods to ensure food safety and minimize potential risks during production and processing. These results reinforce the potential for using organic waste to enhance the productive performance of insects while obtaining high-quality biomass for animal feed. *Phase 3*: This phase evaluated the use of black soldier fly meal as a dietary supplement in diets for cachama blanca. The results revealed that the inclusion of insect meal, as a partial replacement for conventional protein sources such as soybean meal, maintained adequate growth performance and feed conversion efficiency. This underscores the potential of insect meal to reduce reliance on traditional protein sources, which have a significant environmental footprint due to intensive production.

4 Conclusion

Numerous studies demonstrate a positive impact on animal nutrition and agrifood sustainability upon feeding organic waste and/or by-products to insects which are in turn used as animal feed. Selecting appropriate LOCF for insect diets may result in high feed conversion rates, high productive performance, and low toxicological and microbiological risks involved in including insect meal in animal feed. Furthermore, hazards associated with the production of insect meal may be mitigated through good production practices, effective implementation of safety procedures, and the use of analytical tools for tracing insect feed. The results from this research provide key insights for developing public policies that support sustainable value chains, from insect collection and production to commercialization. This project highlights the importance of strengthening international collaborations, such as the Colombia-Germany-Norway partnership, to facilitate knowledge exchange and develop regulatory frameworks promoting insect use in aquaculture and other sectors. As insects play a crucial role in agrifood sustainability, continued research on their species and ecosystem services is essential to unlocking their full potential in the global bioeconomy, advancing circular and resilient production systems.

References

1. Peguero DA et al (2022) Front Sustain Food Syst 5:745894. https://doi.org/10.3389/fsufs.2021.745894
2. Van Zanten HHE, Van Ittersum MK, De Boer IJM (2019) Glob Food Sec 21:18–22. https://doi.org/10.1016/j.gfs.2019.06.003
3. Barragán-Fonseca KB (2023) The potential of insects in sustainable agri-food systems. Book of Abstracts of the 74th Annual Meeting of the European Federation of Animal Science. Lyon, France. https://doi.org/10.3920/978-90-8686-936-7
4. Van Huis A, Oonincx DG (2017) Agron Sustain Dev 37:1–14. https://doi.org/10.1007/s13593-017-0452-8
5. Barragan-Fonseca KB, Dicke M, van Loon JJA (2017) J Insects Food Feed 3:105–120. https://doi.org/10.3920/JIFF2016.0055
6. Barragán-Fonseca KY et al (2022) Trends Plant Sci 27(7):646–654. https://doi.org/10.1016/j.tplants.2022.01.007
7. Barragán-Fonseca KB et al (2013) Small-scale black soldier fly-fish farming: a model with socioeconomic benefits. Anim Front 13:91–101. https://doi.org/10.1093/af/vfad030
8. Dicke M (2018) J Insects Food Feed 4:147–156. https://doi.org/10.3920/JIFF2018.0003
9. Parodi A et al (2021) Waste Manag 126:674–683. https://doi.org/10.1016/j.wasman.2021.04.001
10. Arévalo Arévalo HA et al (2022) Food Control 138:109030. https://doi.org/10.1016/j.foodcont.2022.109030

The Bioeconomy in Colombia from the Perspective of the Colombian Research Center: "AGROSAVIA"

Carolina González-Almario

Highlights

- Bioeconomy represents an economic model that conserves and utilizes the biological resources of agrobiodiversity and manages biomass to generate products, services, and well-being through knowledge, technological development, and innovation

1 Introduction

Bioeconomy, within the framework of a changing global economy, is experiencing rapid growth and encompasses the creation of new pathways in the value chain, the conservation of biodiversity, and the promotion of a circular economy. In Colombia, the International Mission of Wise People has proposed specific goals to boost science, technology, and innovation, highlighting the importance of bioeconomy in reducing climate risk and its potential to represent 10% of GDP [1]. The Green Growth Policy in Colombia also emphasizes the need to sustainably leverage natural capital to diversify the economy and improve efficiency in using natural resources, prioritizing the development of human capital and interinstitutional coordination to promote sustainable projects [2]. Thus, within the previous context, bioeconomy is highlighted as an economy that efficiently manages biodiversity and biomass to generate value-added products, processes, and services through knowledge and innovation [3]. Bioeconomy emerges as a response to social concerns

C. González-Almario (✉)
Corporación Colombiana de Investigación Agropecuaria, AGROSAVIA. Sede Central, Mosquera, Cundimarca, Colombia
e-mail: cgonzaleza@agrosavia.co

A. Meneses-Jácome et al. (eds.), *The Circular Bioeconomy in Industry*, Waste as a Resource, https://doi.org/10.1007/978-3-031-84359-4_15

about resource depletion, land use, and climate change. Productive models have adapted to this trend, seeking to utilize biological resources and conserve biodiversity. This shift toward a biology-based economy involves updating policies, communication, and new institutions to guide behaviors and optimize benefits [4].

In the Colombian context, the Corporación Colombiana de Investigación Agropecuaria (Agrosavia) is a key player in the country's agricultural innovation. Its goal is to sustainably transform the agricultural sector, improving the lives of producers and consumers through knowledge. Considering this objective and the context of bioeconomy in Colombia, Agrosavia has undertaken the task of identifying four strategic lines for addressing this area in the agricultural sector [5]:

- Sustainable agriculture practices: Aim to optimize the use of natural resources and minimize the environmental impact of agricultural production.
- Biodiversity: Focuses on the conservation, characterization, use, and exploitation of agrobiodiversity for the development of new varieties, products, and services.
- Circular economy: Implements strategies for utilizing waste and minimizing waste generation in the agricultural sector.
- Bioproducts: Promotes the development and use of biological-based products for agriculture, industry, and food.

These lines guide the strategies for research, technological development, and innovation (R&D&I), aligning their actions with the needs of the Colombian agricultural sector. Additionally, they help to direct decision-making and actions in terms of R&D&I to effectively address these needs.

2 Material and Methods

Based on the objectives of the Misión de Sabios: Colombia [1], the relevant areas and topics in science, technology, and innovation (STI), and the eight focal points outlined, which include Focus 3: sustainable energy, Focus 4: biotechnology, environment, and bioeconomy, and Focus 8: basic sciences and space, Agrosavia, led by the Department of Institutional Coordination of the Planning and Strategic Direction Directorate, along with the Directorate of Research and Development, the Directorate of Corporate Linkage, researchers, directors of Agrosavia's research centers, and innovation managers and coordinators, established six working groups focused on sustainable agriculture practices, biodiversity, circular economy, bioproducts, PECTIA, and the R&D&I research agenda. These groups aim to define and guide actions and decisions in terms of R&D&I to effectively address and respond to the development of bioeconomy in the country from the agricultural sector, based on research results and contributions from the Corporation in these areas, along with an analysis of political, economic, social, technological, environmental, and legal (PESTAL) components and technological surveillance.

3 Results and Discussion

The bioeconomy guidelines for Agrosavia represent a model of economic development that utilizes biological resources to generate products, services, and well-being. Agrosavia, as an agricultural research center, aims to contribute to the country's bioeconomy through research, development, and innovation. The main objectives include defining the strategic lines and contributions of Agrosavia to the national bioeconomy.

In this context, four key strategic lines are highlighted: (1) sustainable agriculture practices; (2) biodiversity; (3) circular economy, and (4) bioproducts. Sustainable agriculture practices aim to create productive, resilient, and environmentally friendly agricultural systems, utilizing methods such as precision agriculture, agroecology, and organic farming. Biodiversity, essential for agriculture, is addressed through the conservation and sustainable use of ecosystems, species, and genes, with an emphasis on research, the development of bioproducts, and the promotion of sustainable agricultural practices. The circular economy, as an economic model, seeks to reduce, reuse, and recycle resources to minimize environmental impacts, applying strategies such as reducing the use of agricultural inputs, reusing materials, and recycling waste. Agrosavia contributes to this approach through research, technological development, and the promotion of circular economy practices in the agricultural sector. Finally, bioproducts—products derived from biological resources such as biofuels, biomaterials, and pharmaceuticals—are developed and promoted by Agrosavia through research, technological development, and the encouragement of their use in the agricultural sector.

However, there are limitations that hinder the consolidation of strategies at the national and regional levels for better utilization and sustainable use of natural resources. These include a lack of financial resources and allocation criteria, coordination between research agendas and productive sectors, and low effectiveness in institutional planning processes.

In terms of relevance, it is essential to highlight that sustainable production and the responsible use of agrobiodiversity must be strategic objectives to meet the sustainable development goals (SDGs) in environmental, agricultural, food, and rural development aspects. Therefore, national coordination and planning among R&D&I institutions and funding are required to consolidate a coherent strategy that primarily involves areas with high biodiversity. This will enable the development of infrastructure, human resources, and the utilization of natural resources to promote a comprehensive bioeconomy in collaboration with the productive sector. Additionally, sustainable production alternatives in the agricultural sector should be focused on adding value to productive systems, relying on the efficient use of natural resources and the technological development of bioproducts to meet the growing demands of domestic consumption and international markets.

4 Conclusion

In conclusion, the bioeconomy guidelines of Agrosavia represent a crucial tool for the sustainable development of Colombia's agricultural sector, focusing on sustainable agriculture practices, the conservation and sustainable use of biodiversity, the circular economy, and the development of bioproducts. Agrosavia thus aims to contribute to the country's bioeconomy through research, innovation, and capacity-building in these key areas. A coordinated effort between the public sector, private sector, and academia is required to overcome limitations and seize the opportunities that bioeconomy offers to the country's development.

References

1. Misión de Sabios: Colombia (2019) Gobierno de Colombia. https://minciencias.gov.co/sites/default/files/libro_mision_de_sabios_digital_1_2_0.pdf (In Spanish)
2. Departamento Nacional de Planeación (2018) Política de Crecimiento Verde (Documento Conpes 3934). https://bit.ly/3tDkfa4 (In Spanish)
3. Ministerio de Ciencia Tecnología e Innovación (2020) Bioeconomía para una Colombia Potencia viva y diversa. https://minciencias.gov.co/sites/default/files/upload/paginas/bioeconomia_para_un_crecimiento_sostenible-qm_print.pdf (In Spanish)
4. ANDI (2023) Lineamientos para la Bioeconomía en el Tolima. https://www.andi.com.co/Uploads/Lineamientos%20para%20la%20Bioeconomi%CC%81a%20en%20el%20Tolima.pdf. consultado en Febrero 2024. (In Spanish)
5. Corporación Colombiana de Investigación Agropecuaria, Mosquera (2021) Bioeconomía: lineamientos corporativos. Repositorio Institucional Agrosavia. http://hdl.handle.net/20.500.12324/36909 (In Spanish)

Circular Bioeconomy to Lead an Enhanced Strategy for Sustainable Aquaculture: CiBELESα

Tatiana C. Guarín C., Lorena Durán-Ortiz, Sebastián Muñoz-Chinome, Astrid Hernandéz-Rincón, Yohana Castro-Hernández, and Alexander Meneses-Jácome

Highlights

- To achieve sustainable aquaculture, it is crucial to improve technologies and manage production intensity to balance environmental impacts effectively.
- The evaluation of "Eutrophication Potential" must be included in the life cycle assessment (LCA) analysis.
- CiBELESα focused on innovative biotechnology for microorganism production, bio-based product creation, and analyzing circular bioeconomy opportunities/constraints.

1 Introduction

Aquaculture as an economic practice has grown in recent years worldwide and Colombia has not been the exception, including among others artisanal and small community initiatives of local fishermen and farmers in the main river basins of the

T. C. Guarín C.
Grupo de Investigación en Recursos Hídricos y Saneamiento Ambiental - GPH, Escuela de Ingeniería Civil, Universidad Industrial de Santander-UIS, Bucaramanga, Colombia

L. Durán-Ortiz · S. Muñoz-Chinome · A. Hernandéz-Rincón · Y. Castro-Hernández · A. Meneses-Jácome (✉)
Circular Bioeconomy Research Center (ε-BiO): Biotechnology and Environment Group, Universidad Autónoma de Bucaramanga (UNAB), Bucaramanga, Colombia
e-mail: ameneses2@unab.edu.co

A. Meneses-Jácome et al. (eds.), *The Circular Bioeconomy in Industry*, Waste as a Resource, https://doi.org/10.1007/978-3-031-84359-4_16

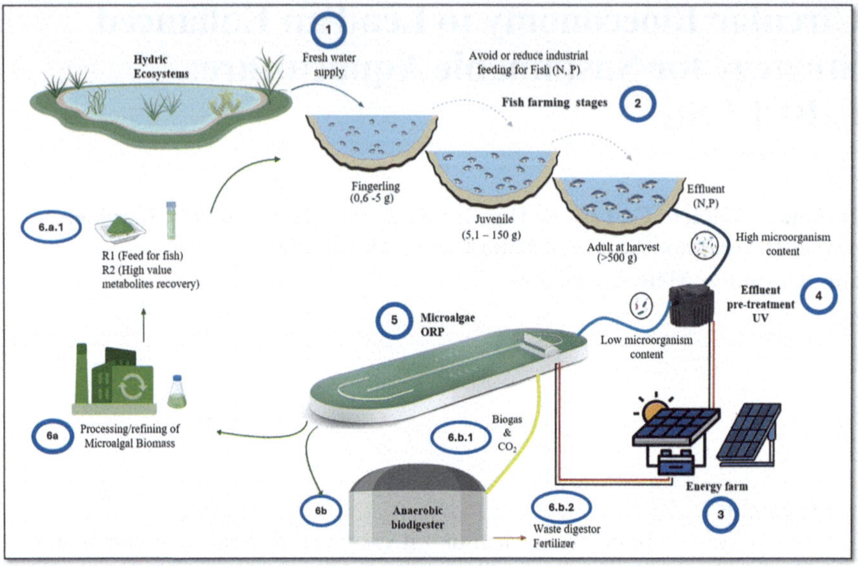

Fig. 1 CiBELESα's project processes schema

country [1]. This short communication documents the progress related to the CiBELESα's project, whose main objective is to improve inland aquaculture systems in Colombia by implementing the cultivation of microalgae as a technological option for nutrient recovery from the effluents generated by this activity. In Colombia, artisanal inland freshwater aquaculture carried out in neighboring areas of the Magdalena River basin and its wetlands discharges nutrient-rich effluents, threatening the sustainability of these protected ecosystems. CiBELESα promotes the production of feed derived from microalgae while simultaneously purifying water and promoting strategies for reuse and protection of this resource, offering long-term protection for hydric ecosystems and a diversified income for farmers. Furthermore, an "energy farm" with solar cells was considered to ensure a self-sufficient renewable energy supply for this system and a low carbon footprint for the full system operation. Figure 1 summarizes the circular bioeconomy concept suggested by the CiBELESα project.

2 Material and Methods

To manage the project, the proposed methodology utilizes a "Work Breakdown Structure" (WBS) approach. In this way, the project is defined by means of three technical tasks or work plans (WPs), which were decomposed into activities with specific deliverables. The estimation of the duration of the development and execution of each WP was scheduled within the project's timeframe (36 months). The WPs are described below:

2.1 WP1: Baseline Study Aquaculture Systems

It covers to establish a technical database and inventory about the technical environmental aspects of local fish farming technologies, energy flows, and services of the bioeconomy model, the socio-economic role of the postculture sector, including productivity, cost analysis, and identification of social gaps, and a review of local environmental regulations for aquaculture.

2.2 WP2: Microalgae Cultivation Systems and Algal Biomass Essays as Feed

It was focused on the implementation of the effluent conditioning system and the implementation of a pilot Open-Raceway Pond (ORP) bioreactor for microalgae cultivation and harvesting. It includes, as a complement, the implementation of a system of photovoltaic energy to drive the integral operation of this system. Then, the experimental component included advances related to the conceptual designs of the pilot plant for microalgae cultivation and its complementary components. Within the complementary components, conceptual designs of pretreatment systems and the photovoltaic farm were developed to ensure sustainable energy for the system.

Before sowing the microalgae in the open-pond raceway (OPR) pilot bioreactor, an isolation and bioaugmentation study of the selected strain was carried out at the laboratory level. Indeed, measurements of total suspended solids (TSS) were carried out using the standard method (SM 2540D) to evaluate the biomass production of *platensis* in glass photobioreactors under controlled laboratory conditions. The cultivation process was conducted with a 12-h light and dark cycle, bubble column agitation, an average temperature of 25 °C, and the use of Zarrouk's standard culture medium supplemented with an inoculum equivalent to 10% of the effective volume of the bioreactor. Monitoring was performed every 72 h over a period of 21 days, in triplicate, and using the same Zarrouk culture medium as a control.

In addition, *Spirulina* sp., in dry and powdered form, was incorporated into the diet of *Oreochromis* sp. (red tilapia) fry at three different concentrations (0%, 10%, and 25%), replacing a part of the protein component of the commercial feed usually applied by the local fish farmers. The experiment was conducted over 42 days in three aquariums, each containing 38 fry with an average weight of 1 g. Morphometric characteristics such as body weight and total length were recorded, and water quality was monitored, ensuring that physicochemical parameters remained within an optimal range for fry growth. This design allowed for the evaluation of the microalgal biomass inclusion in their diet.

2.3 WP3: Sustainability Assessment and Economic Analysis

It relates to carrying out a life cycle analysis (LCA) and economic feasibility study to determine the applicability of the model on a real scale, considering regulatory restrictions to assess its true potential as a circularity concept. The methodological choices for the LCA application have been completed, and currently, the inventory data is being consolidated. The next stage corresponds to the assessment of the mid-point indicators particularly those corresponding to eutrophication potential, global warming potential, acidification potential, and cumulative energy demand being proposed as sensitive for this kind of process. Furthermore, LCA results will be compared with those being obtained from the twin project of CiBELES being carried out in Germany by the "Microalgae biotechnology development group" of the Fraunhofer Institute in Stuttgart. This comparison aims to establish convergent and differentiating points between land-based fish farming carried out in two totally opposite contexts. Finally, this methodological aspect also involves the commitment of the working team to participate and cooperate with the local community to influence the visibility of the project through participatory, educational, and communication plans, which have been followed to improve and facilitate research and dissemination of results to all stakeholders by means the launch of the project website (www.proyectocibeles.com).

3 Results and Discussion

This contribution is specifically focused on results from WP2, as follows:

3.1 OPR Bioreactor Pilot System and Solar Farm

CiBELESα project installed an Open-Pond Raceway (ORP) bioreactor of 8 m^3 capacity, 30 cm depth, and a paddlewheel at 4–15 RPM. The bioreactor was installed in a fish farm of red tilapia (*Oreochromis sp.*) located in the rural area of the Chimichagua's municipality (coordinates: 9°14′90.9′'N and 73° 1′50.8′'W) in the immediate vicinity of the Zapatosa Swamp, one of the main wetlands formed by the Magdalena River in its complete journey through Colombian territory from the south to north.

To design the solar farm as a source of renewable energy to operate fully the OPR bioreactor and the auxiliary equipment, the local climatic conditions were analyzed from reported data on the website https://es.weatherspark.com/. The minimum irradiance and daily insolation considered for the solar farm design corresponded to the most unfavorable scenario presented during the month of May (757.2 W/m^2 at 12 m; 5.34 kWh/m^2), establishing 5.34 peak sun hours. The total system load includes LED lamps perimeter lighting, a UV filter, aquarium pumps, and a bioreactor drive

Fig. 2 CiBELESα Open-Raceway Pond reactor

system, all in full-time use, with a power and energy demand of 220 W and 3474 Wh, respectively.

The installed photovoltaic solar system consists of two (2) *Q-cells* PV modules, reference *Q.PEAK DUO XL*-G9.3, of 455 Wp, connected in series; a low-frequency inverter charger, *Must-Solar PV3000 LMPK*, with 1 kW power, 120 Vac to 24 Vdc; and two (2) gel batteries, *GreenPoint 6-CNFJ-250*, of 12 V and 250 Ah (one (1) autonomy day). A voltage regulator was installed to protect the inverter. The drive force system of the bioreactor required a transformer and frequency inverter before the electric motor, an *IE3* 0.37 kW 1/2 HP three-phase induction electric motor with super high efficiency. Figure 2 shows the implementation of the bioreactor, aquariums for the controlled growth of tilapia fry partially fed with *Spirulina* biomass, and the installation of solar cells.

3.2 Select Relevant Laboratory Results

An average value of 230 mg/L of total suspended solids (TSS) was achieved at the end of the 18-day cultivation period, with average variations between replicates remaining within acceptable ranges for controlled laboratory conditions. The observed trend in cultivation suggests a continuous increase in biomass production over time, with a projection to exceed 500 mg/L in later stages under improved conditions (see Fig. 3). These results provided a solid foundation for developing guidelines for future applications in field systems and serve as a reference for scaling up in OPR-type bioreactors. Additionally, they enable the evaluation of the feasibility of growing *Spirulina platensis* by using wastewater from fish-fattening ponds as a substrate, reinforcing its potential in circular bioeconomy systems focused on sustainable aquaculture.

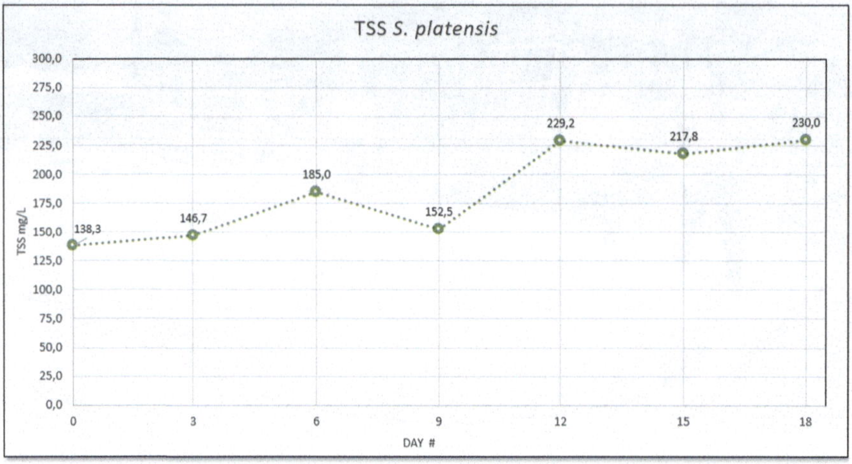

Fig. 3 *Spirulina platensis* biomass production in laboratory

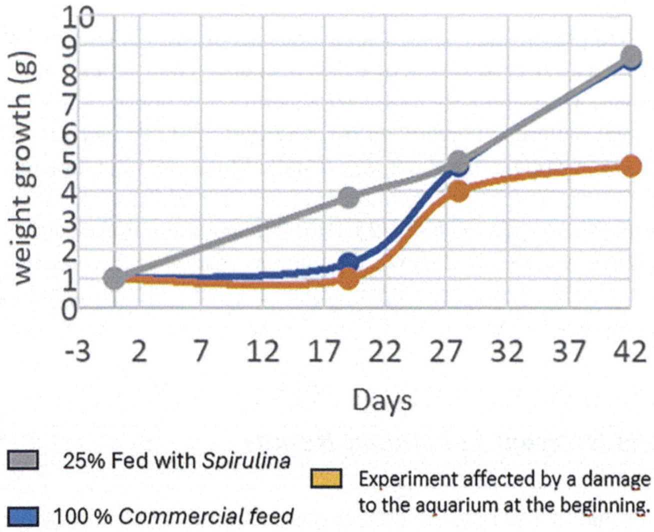

Fig. 4 Growth rate per fish (red tilapia)

3.3 *Spirulina* Biomass as Feed for Juvenile Fish (Red Tilapia Fry)

The results have shown that the inclusion of *Spirulina* sp. in the diet of *Oreochromis* sp. fry affects positively their growth and feed efficiency. The control group (0% Spirulina feed, blue line in Fig. 4) had a weight gain of 7.48 g, while the groups with *Spirulina* sp. (25%, gray line) had gains of 7.61 g, but with a higher gain in the first

20 days. Although the growth rate was lightly highest in the control group (17.81%), the groups with *Spirulina* sp. showed better efficiency and feed conversion, with the 25% *Spirulina* sp. group having a favorable weight–length ratio and a feed conversion index of 3.08 g/g. P and N concentrations were about 0.5–1.5 kg P/100 kg of fish, with an N/P ratio of 4–6.

4 Conclusions

- As a valuable finding regarding local artisanal aquaculture practices, it is clear that they must be strengthened through training and education. This will improve sustainable technologies within the fishing communities, enabling them to adopt less harmful practices and benefit of innovative solutions.
- The increasing demand for protein-rich compounds underscores the promising protein potential of microalgae as feed for fish, which could enhance fry growth rates and facilitate commercialization.
- The production of microalgae in environments that generate lower negative environmental impacts and are more sustainable, such as the system proposed in CiBELES, is an attractive pathway and warrants attention. Managing and controlling effluents with a focus on nutrient recovery and their reuse for microalgae growth adds valuable benefits and diversifies the incomes of local populations. In this way, this study strengthens previous evidence on how coupling microalgae cultures with fish farming systems contributes to more sustainable performance. [2–5]
- CiBELESα is an ongoing project that will continue following its clear, concise, and well-structured methodology to achieve its established objectives and goals.

Acknowledgments The authors would like to thank the Ministry of Science, Technology and Innovation of Colombia (MinCiencias) and the Federal Ministry of Education and Research of Germany (BMBF), which through the call for proposals "Bioeconomy International 2020" provided budgetary support to the CiBELESα project (Contract 80740-446-2021). The authors also thank the research team of the Fraunhofer Institute IGB of Germany and its collaboration with the research group of the Universidad Autónoma de Bucaramanga in the frame of the twin CiBELESα projects.

References

1. Carrera-Quintana SC, Gentile P, Girón-Hernández J (2022) An overview on the aquaculture development in Colombia: current status, opportunities and challenges. Aquaculture 561:738583. https://doi.org/10.1016/j.aquaculture.2022.738583
2. Tom AP, Jayakumar JS, Biju M et al (2021) Aquaculture wastewater treatment technologies and their sustainability: a review. Energy Nexus 4:100022. https://doi.org/10.1016/j.nexus.2021.100022

3. Ribeiro-Cargnin JM, João JJ (2021) Removal of nutrients from aquaculture residual water: a review. Ambiente e Agua 16:1–20. https://doi.org/10.4136/ambi-agua.2747
4. Ahmad AL, Chin JY, Mohd Harun MHZ, Low SC (2022) Environmental impacts and imperative technologies towards sustainable treatment of aquaculture wastewater: a review. J Water Process Eng 46:102553. https://doi.org/10.1016/j.jwpe.2021.102553
5. Ghamkhar R, Boxman SE, Main KL et al (2021) Life cycle assessment of aquaculture systems: does burden shifting occur with an increase in production intensity? Aquac Eng 92:102130. https://doi.org/10.1016/j.aquaeng.2020.102130

Food Security and Sustainability Based on Alternative Proteins from Insects in the Santander Region (Colombia): A Perspective

Yudy Mariana Alfaro-Wisaquillo

1 Scientific Perspective

By 2050, the consumption of animal products is expected to increase between 60% and 70%. For this reason, exploring new sources of protein becomes an overriding necessity for the agricultural sector, and the FAO (2009) suggests the use of insects in human and animal nutrition as an alternative nutrient source due to their abundance and nutritional characteristics, with a possible lower environmental impact than usual animal protein, which would need to be specifically proved in each specific context.

For instance, in the region of Santander (Colombia), the consumption of a kind of ant, named "hormiga culona" or *Atta laevigata,* has been a well-known ancient practice, even before the arrival of the Spanish colonists in South America. This practice, in the current levels of local and commercial consumption, has exposed this species to a notorious population decline, pointing out that further research and more knowledge are required to prevent environmental and social impacts associated with expanding the exploitation of this kind of alternative protein to develop a sustainable food supply chain. This includes ecosystem protection measures, sustainable management programs, and the creation of food and gastronomic products, as well as the understanding of the complex links between nature and culture; it means and requires the adoption of a holistic approach to identify synergies and areas of opportunity for the adoption and promotion of alternative proteins, such as those provided by certain insects, within the framework of sustainable food systems.

The promotion and adoption of alternative proteins, such as *Atta laevigata* or other native insects with nutritional potential, as part of sustainable food systems, can generate significant economic, social, and environmental benefits in the

Y. M. Alfaro-Wisaquillo (✉)
Universidad Autónoma de Bucaramanga—UNAB, Bucaramanga, Colombia
e-mail: yalfaro321@unab.edu.co

A. Meneses-Jácome et al. (eds.), *The Circular Bioeconomy in Industry*, Waste as a Resource, https://doi.org/10.1007/978-3-031-84359-4_17

Department of Santander. This hypothesis is based on the premise that diversifying protein sources can contribute to food security, natural resource conservation, and local economic development. However, comprehensive conservation strategies must be designed, along with guidelines for the development of sustainable new food products. These outputs would be the result of a multidisciplinary research approach, combining aspects of ecology, economics, gastronomy, and community development.

Regarding a possible research-innovation strategy in this field, the exhaustive analysis of the local socioeconomic context would include the widest variety of stakeholders, rural communities, local authorities, experts in ecology and gastronomy, as well as potential marketers and consumers of these alternative proteins. In the next level, detailed information on the global production and consumption of alternative proteins, for instance, quantity produced, market trends, and consumption patterns at the global and national levels, must be analyzed. Likewise, the environmental footprint of these new food supply chains would be compared to conventional animal proteins, within the framework and scope of life cycle assessment methods. These results are expected to inform and support public policies and business decisions aimed at endorsing the sustainability of insect-derived food.

In conclusion, the expected impacts of the successful implementation of a sustainable commercial exploitation of native insects as a source of protein must cover various key aspects of society, economy, and environment. A positive impact on biodiversity conservation and the health of local ecosystems is mandatory and the adoption of alternative proteins instead of conventional ones of animal origin should demonstrate its capacity to reduce the pressure on natural resources and other associated environmental impacts to food supply. In addition, by means of the diversification of the local gastronomic offer, rural communities could diversify and improve their incomes and recover the regional cultural and culinary identity by strengthening community ties. From here, it is expected that this line of research and innovation, for the use of the biomass of insects and the ecosystems where they grow, will be incorporated in a more active way in the financing schemes of the local and national bioeconomy.

Part III
Other Bioeconomy and General Sustainability Topics

Exploration of Life Cycle Assessment as Driver of Circular Economy Strategies in the Drinking Water Sector: A Literature Outlook

Fabián E. González-Esteban and Alexander Meneses-Jácome

Highlights
- Life cycle assessment (LCA) is presented as a foundational tool to promote circular economy practices in drinking water treatment systems.
- The presence of "circular economy"associated with LCA applications for the drinking water sector unfolds a niche theme of scientific interest.

1 Introduction

The circular bioeconomy stands as an emerging frontier in the quest for environmental sustainability, particularly within the drinking water treatment sector. Despite its critical importance for public health and environmental protection, the water purification infrastructure—including aqueducts, capture facilities, and distribution networks—has received less attention in environmental assessment using the life cycle assessment (LCA) methodology than that of the sanitation sector. This

F. E. González-Esteban
Circular Bioeconomy Research Center (ε-BiO): Biotechnology and Environment Group, Universidad Autónoma de Bucaramanga (UNAB), Bucaramanga, Colombia

Centro de investigación del agua CENIAGUA, Acueducto Metropolitano de Bucaramanga amb S.A E.S.P, Bucaramanga, Colombia
e-mail: fgonzalez830@unab.edu.co

A. Meneses-Jácome (✉)
Circular Bioeconomy Research Center (ε-BiO): Biotechnology and Environment Group, Universidad Autónoma de Bucaramanga (UNAB), Bucaramanga, Colombia
e-mail: ameneses2@unab.edu.co

A. Meneses-Jácome et al. (eds.), *The Circular Bioeconomy in Industry*, Waste as a Resource, https://doi.org/10.1007/978-3-031-84359-4_18

disparity reflects not only the differences in technological functions and environmental implications between the two sectors but also a significant opportunity to expand our understanding and application of LCA in the sustainable management of drinking water systems. An exploration of bibliometrics and thematic distribution in the scientific literature reveals a nascent but growing interest in integrating the principles of the circular economy into drinking water treatment. This interest outlines the movement toward practices that not only reduce environmental impact but also promote resource regeneration, thus aligning with global principles of sustainability and environmental protection. This research focuses on the application of LCA as a fundamental tool to establish an environmental matrix and baseline sustainability conditions and highlights its crucial role in advancing sustainable practices in the water sector.

2 Material and Methods

In an examination of the application of life cycle assessment (LCA) as a tool to assess the environmental performance of water purification infrastructure, which includes aqueducts, capture facilities, and distribution networks, it is observed that this area has received much less attention compared to the sanitation sector, encompassing sewers and wastewater treatment plants. This disparity may be attributed to the distinct technological functions of both systems and the more immediate and complex environmental implications associated with treating wastewater safely. A comparative analysis of bibliometric indicators reinforces this observation. Within the Web of Science database, the search for articles using the query "(All Fields: (('wastewater') AND ('Life Cycle Assessment' OR LCA))" yielded 2056 articles. In contrast, the search with "(All Fields: (('Drinking water') AND ('Life Cycle Assessment' OR LCA))" found 317 articles, and a further refined search including "(All Fields: (('Drinking water treatment' OR 'drinking water system') AND ('Life Cycle Assessment' OR LCA))" located just 48 articles, without imposing temporal restrictions. These findings, being summarized in Fig. 1, suggest a relative interest among the scientific community in the environmental assessment of drinking water infrastructure through LCA and point out it as the driver to move forward to circular economy appraisals in this field.

The thematic distribution map hints at an intriguing, albeit underexplored, nexus within the realm of drinking water research: the role of the circular economy. Despite its lower centrality in the literature compared to core themes such as "drinking water" and "life cycle assessment," the presence of "circular economy"within the niche themes indicates an emergent scholarly interest. This interest is further corroborated by the bibliometric analysis, which, although revealing a modest number of articles, points to a substantive intersection between the valorization of residuals and sludge from water treatment processes for human consumption and the pursuit of a circular economy in these facilities.

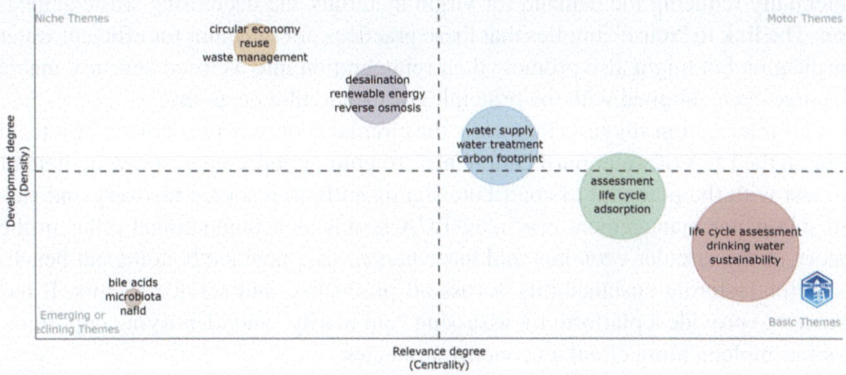

Fig. 1 Thematic distribution map generated with Bibliometrix for the search query "(All Fields: (('Drinking Water') AND ('Life Cycle Assessment' OR LCA))" run on the Web of Science database

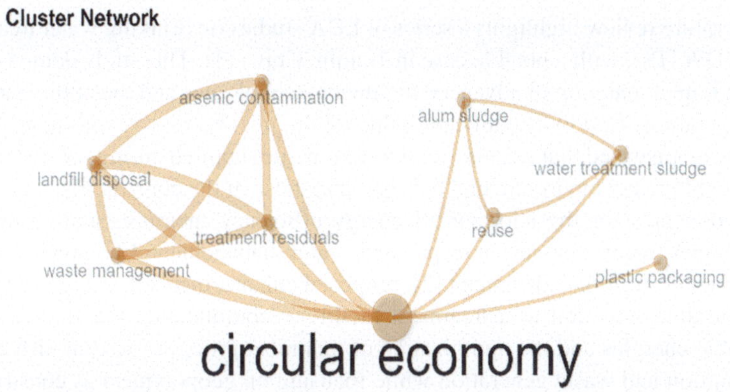

Fig. 2 Cluster Network, generated with Bibliometrix, delineates the scope of 48 articles obtained through the Web of Science database using the search query "(All Fields: (('Drinking water treatment' OR 'drinking water system') AND ('Life Cycle Assessment' OR LCA))"

On the other hand, Fig. 2 illustrates the Cluster Network of the circular economy within the framework of LCA in the potabilization sector, highlighting how circular economy practices intertwine with environmental sustainability in water treatment. On one side, elements such as "arsenic contamination," "landfill disposal," "waste management," and "treatment residuals" might suggest a need to develop effective strategies for secure disposal and waste minimization coming from water purification facilities, which is crucial for mitigating environmental impact at landfills and soils. On the other side, the relationship between "alum sludge," "water treatment sludge," and "reuse" sheds light on the commitment to valorizing the byproducts of water treatment. Proper management and reuse of these sludges could be key to transforming these residues into valuable resources, applicable in various areas, potentially reducing the demand for virgin materials and decreasing waste generation. The link to "reuse" implies that these practices not only aim for efficient water purification but might also promote their reintegration into a closed and sustainable resource cycle, aligned with the principles of the circular economy.

This intersection suggests that while the circular economy may not yet be a focal point in the LCA of water purification infrastructure, it represents a growing field of interest with the potential to contribute significantly to resource recovery and sustainable water management practices. LCA stands as a foundational pillar in the theory of the circular economy and has emerged as a new methodological benchmark for fostering sustainability across all productive and service sectors. It has evolved to provide a platform for assessing "circularity" and identifying opportunities for implementing circular economy strategies.

3 Results and Discussion

The literature review highlights a series of LCA studies on drinking water treatment plants (DWTPs), with a notable case in Tianjin, China [1]. This study demonstrated that the implementation of advanced treatment technologies and the optimization of plant operations lead to a significant reduction in greenhouse gas emissions. It particularly emphasized that circular economy strategies applied during the operational phase yield greater environmental benefits than those of the construction phase. The study advocates for the adoption of energy efficiency measures and sustainable technologies as cornerstones of responsible water management. Echoing these findings, research from Iasi, Romania [2], proposed enhancements in energy efficiency and emission reduction at a treatment facility, recommending the integration of renewable energies and the optimization of treatment processes to diminish energy consumption and waste generation while focusing on geopolymers as construction materials with minimal environmental impact. Correspondingly, a study in Morocco [3] showed how sludge from water treatment processes could be incorporated into the production of geopolymers, providing a sustainable alternative to conventional Portland cement and improving material properties as well as reducing the carbon footprint in construction. This discovery underscores the significance of reusing byproducts as a key circular practice within the construction industry. Meanwhile, in Canada [4], a human health risk assessment (HRA) approach was employed to select water treatment technologies, highlighting the need for effective and environmentally responsible purification that aligns water quality with the ecological integrity of the process. Collectively, these studies demonstrate that the implementation of circular economy strategies in drinking water treatment is shaping up as an effective path toward more sustainable and efficient systems. Life cycle assessment emerges as a critical tool to encourage practices that not only minimize environmental impact but also enhance resource regeneration, in harmony with global principles of sustainability and environmental protection.

4 Conclusion

Life cycle assessment (LCA), when focused on an approach toward circularity evaluations of drinking water systems, gains increased added value and scientific as well as sectoral interest. This heightened significance stems from the opportunities to achieve more sustainable drinking water infrastructure from a circularity perspective. These opportunities include understanding and enhancing local conditions to safeguard water reserves feeding the system, selecting or improving the technology used in aqueducts with respect to the rational use of chemical inputs and energy, reducing factors contributing to climate change, and minimizing sludge generation. Additionally, the implementation of waste disposal practices or valorization/reuse strategies that lessen the impact on ecosystems and human health compared to traditional practices, such as landfilling or returning to water bodies, is crucial. This comprehensive approach not only aims at enhancing the efficiency and sustainability of drinking water systems but also at ensuring the preservation and responsible management of vital water resources.

Within this framework, the Colombian Government has identified a significant opportunity to align its policies with circular economy principles in the drinking water treatment sector [5], as a core concept of sustainability within its operational activities and the modernization of its infrastructure. However, to implement and contribute to the proper governance of this sectorial policy, the design or formulation of circularity indicators tailored for this purpose seems to hinge on the robust foundation provided by LCA, marking its pivotal role in advancing sustainable practices and its capacity to couple with complementary theories and methods to incorporate other sustainability aspects at the ecosystemic, economic, and social dimensions. Indeed, this issue has begun to be understood in recent years and is gradually incorporated as part of pioneering research initiatives to structure modern environmental regulations, including circularity criteria as sustainability notions, and to provide water companies with tools to confront future challenges in this domain.

Acknowledgments The authors would like to thank the operating company of the Metropolitan Aqueduct of Bucaramanga, amb S.A E.S.P., for funding the doctoral research of Mr. Fabian Gonzalez.

References

1. Zhang P et al (2024) Ecotoxicol Environ Saf 272(October 2023):116043. https://doi.org/10.1016/j.ecoenv.2024.116043
2. Bârjoveanu G et al (2019) Environ Eng Manag J 18(2):513–522. https://doi.org/10.30638/eemj.2019.048
3. Bourzik O et al (2023) Environ Sci Pollut Resear Int 30:118989–119000. https://doi.org/10.1007/s11356-023-30736-w
4. Hu G et al (2020) J Clean Prod 267:121980. https://doi.org/10.1016/j.jclepro.2020.121980
5. CONPES (2020) Documento CONPES 4004: Economía Circular en la Gestión de los Servicios de Agua Potable y Manejo de Aguas Residuales. Consejo Nacional de Política Económica y Social de Colombia - Departamento Nacional de Planeación, Bogotá, p 64

Using Animation Technologies to Improve the Interrelation Between Bird's Conservation and Nature Tourism

Daniel Tello and Olga Ocampo

1 Introduction

The sustainable tourism paradigm aims to balance environmental sustainability, economic development, and social welfare, particularly in local communities engaged in ecotourism. However, increased tourism can negatively impact fragile ecosystems and biodiversity, such as the conservation of high-value endemic and migratory birds in emerging green tourism destinations like "Los Santos" in Colombia, as well as other regions in this country, the richest in bird species [1]. While existing research examines tourism's effects on natural spaces, more interdisciplinary studies are needed to find unconventional solutions that involve active community participation [2]. This qualitative, interdisciplinary research-creation study from the arts and humanities perspective argues that animation technologies have a significant potential to raise awareness about bird vulnerability and the need for protection. This study aims to establish conceptual and methodological foundations for using 4.0 animation pieces for conservation, incorporating community perspectives and local knowledge to strengthen the sustainability of this type of local bioeconomy [3].

D. Tello (✉)
Universidad Autónoma de Bucaramanga—UNAB, Bucaramanga, Colombia
e-mail: dtello@unab.edu.co

O. Ocampo
Universidad Autónoma de Manizales—UAM, Manizales, Colombia
e-mail: olocampo@autonoma.edu.co

A. Meneses-Jácome et al. (eds.), *The Circular Bioeconomy in Industry*, Waste as a Resource, https://doi.org/10.1007/978-3-031-84359-4_19

2 Material and Methods

"Avitourism," is a form of nature tourism focused on bird observation and identification in natural habitats, which serves as leisure and recreation for both novice and expert observers, offering aesthetic and artistic experiences through illustration, photography, video, or audio capture for bird registration and classification. According to the Colombia Ministry of Commerce, countries such as Australia, England, and the United States excel in tourist infrastructure, accessibility, and promotion of birdwatching, attracting numerous visitors and generating substantial economic revenues [4]. Despite lacking the same level of infrastructure, Latin America and Colombia, with their mega biological diversity, are appealing alternatives for birdwatchers. Colombia, hosting about 20% of all bird species globally, is a significant destination for "avitourists" [5]. Table 1 details the number and characteristics of these birds.

Among Colombia's regions, Santander has not yet fully developed avitourism. Despite recording 240 visitors in 2023, precise data on local and foreign birdwatching tourists is unavailable. However, information on bird species and classifications, shown in Table 2, highlights Santander and, particularly the municipality of "Los Santos" as an interest point for this local bioeconomy sector [6].

The municipality of "Los Santos" in Santander (Colombia) covers 467 km^2, has an average temperature of 21 °C and a population of about 12,772 inhabitants, and is pointed out as an ideal location to promote birdwatching by the Ministry of Commerce, Industry, and Tourism, which stresses the importance of considering economic, social, and environmental impacts on visitors, the tourism industry, and local communities. Balancing bird conservation, community needs, and tourism quality is essential. New technologies can help by integrating ethics and artistic expression to focus on environmental care and bird conservation, avoiding prioritizing human needs over ecosystems [7].

This project uses "Los Santos" as a case study, investigating its context, bird species, and ecotourism potential. It explores animation, technology, circular tourism, and sustainable development to create an animation model for bird conservation and nature tourism. The study employs a research methodology based on artistic practice with an ethnographic approach, using photography and ornithology to design an animation artifact, critically addressing the relationship between bird conservation and nature tourism [8]. It highlights the interrelation of green tourism activities with ecosystem sustainability, local communities, and biological asset care in Los Santos.

Table 1 Number of birds in Colombia

Description	# Species of birds
Total, avian specimens	1921
Endemic	79
Migratory, boreal, and austral species	275

Source: MinComercio 2021 (Colombian Ministry of Commerce, Industry, and Tourism, 2021)

Table 2 Number of birds in Santander

Description	# Species of birds
Total, avian specimens	935
Endemic	28
Migratory	102
Exotic	3
National threat	31
Global threat	23

Source: MinComercio 2021 (Colombian Ministry of Commerce, Industry, and Tourism, 2021)

Fig. 1 Phases and objectives

The research includes six phases, shown in Fig. 1, but this document details phases 1, 2, and 3, identifying key animation experiences in avifauna preservation and nature tourism to highlight the most representative bird species of Los Santos and their cultural significance.

Phases 1, 2, and 3 of the study use techniques such as Information Gathering, Unstructured Observation, and Participant Observation, with data collected through field trips, journals, and audiovisual recordings. These inputs aid in creating animation pieces to be tested in workshops and focus groups. As depicted in Fig. 2, essential study elements include a monocular or spotting scope with around 1000 mm focal length for birdwatching, enabling observation from a distance without disturbing the birds. Combining the monocular with a mobile phone holder allows for artisanal image and video capture, offering a unique aesthetic compared to digital cameras. These elements support the design, creation, and testing of the artifact in phases 5, 6, and 7.

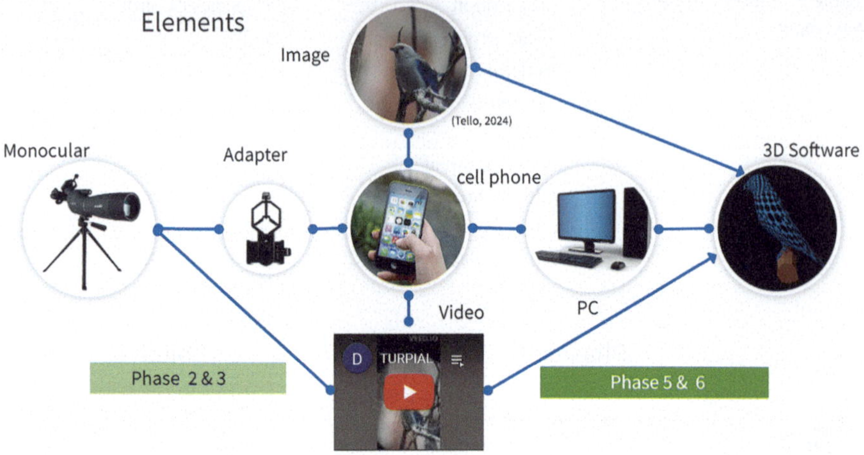

Fig. 2 Elements

3 Results and Discussion

This section details the arguments and findings, proposing new theoretical and applied guidelines to help sustainability actors promote cultural and environmental values, generating new knowledge and actions in nature, art, culture, and birdwatching [9]. In phases 1 and 2, key elements such as bird vulnerability to tourism and poor practices by birdwatchers were identified. Local perspectives on value appropriation, represented in stories and symbols, highlighted themes and areas for improvement for the animation pieces in phases 5 and 6. Feedback from key audiences and communities was considered, integrating scientific, cultural, and social knowledge to inform conservation strategies and promote sustainable tourism and birdwatching [10].

4 Conclusion

The study in its three phases yielded the following conclusions:

- It is necessary to work with the community for an effective appropriation of avian assets.
- Capturing images with a monocular is complex; however, it is an economical tool and adds a craft aesthetic to the design of the study.
- Although there is skepticism regarding research + creation, it is possible to integrate a methodology grounded in artistic practice in the field of bird sustainability and conservation.

References

1. Bertoni M (2008) Turismo sostenible: su interpretación y alcance operativo. Cuadernos de Geografía: Revista Colombiana de Geografía 17:155–163
2. Francisco. (2015) LAUDATO SI. Iglesia Católica
3. MinCiencias (2024) Investigación+Creación. Obtenido de Gobierno de Colombia. https://minciencias.gov.co/investigacion-creacion/que-es-ic
4. Cabiedes D (2019) impacto ambiental del ecoturismo y las normas de sostenibilidad ambiental en las Áreas Protegidas de Colombia. ECOVIDA 9(1):119–140
5. Perdomo O, Salazar P, Fernández L (2018) Avifauna local: una herramienta para la conservación, el ecoturismo y la educación ambiental. Ciencia en Desarrollo 9(2):17–34
6. Sánchez A, Varona F, Avilas J, Ochoa M (2023) Comportamiento del turismo de naturaleza en la gestión de destinos turísticos. Ciencias Holguín 29(2):1–12
7. Millán J (2022) Ilustrar y animar para conservar: Sobre aves endémicas en los humedales de Bogotá. Universidad Pedagógica Nacional, Bogotá
8. Tello D (2018) Realidad mixta/híbrida para la enseñanza de animación. In: Dal Farra R (ed) Diseño y Creación: Festival Internacional de la Imagen, vol 17. Universidad de Caldas, Manizales, pp 407–412
9. Ecologistas en Acción (2020) Impacto del turismo en los espacios naturales y rurales. Obtenido de. https://www.ecologistasenaccion.org/wp-content/uploads/2020/09/informe-turismo-rural.pdf
10. García I, Pompa S, López Á (2017) Ecoturismo como herramienta para promover el empoderamiento: el caso del Ejido San Francisco en el Área de Protección de Flora y Fauna Sierra de Álvarez. San Luis Potosí El periplo Sustentable 32:1–35

Advances and Perspectives in Processing Techniques and Artificial Intelligence Methods for Biomedical Signal Analysis

Hernando González, Carlos Julio Arizmendi, and Beatriz F. Giraldo

Highlights

- Signal processing techniques are transforming biomedical signal analysis, providing new tools for disease detection and treatment optimization.
- The integration of machine learning and deep learning algorithms in the interpretation of biomedical signals is advancing, improving the accuracy and efficiency of clinical diagnoses.
- The potential impact of artificial intelligence on medical sustainability is highlighted, emphasizing the role of this technology in ensuring the quality and safety of patient care.

1 Introduction

Biomedical applications and data science for biointelligence were considered within the first national bioeconomy approach by the Colombian Mission of the Wise in 2019 (Mincieincias 2019). For instance, research on biomedical equipment has received a considerable boost from then, and signal analysis by means of artificial intelligence (AI) media has been a specific field of research and innovation on this

H. González (✉) · C. J. Arizmendi
Universidad Autónoma de Bucaramanga—UNAB, Bucaramanga, Colombia
e-mail: hgonzalez7@unab.edu.co; carizmendi@unab.edu.co

B. F. Giraldo
Universidad Autónoma de Bucaramanga—UNAB, Bucaramanga, Colombia

Universidad Politécnica de Cataluña—UPC, Cataluña, Spain
e-mail: beatriz.giraldo@upc.edu

A. Meneses-Jácome et al. (eds.), *The Circular Bioeconomy in Industry*, Waste as a Resource, https://doi.org/10.1007/978-3-031-84359-4_20

front. Biomedical signal analysis has undergone an evolution driven by the integration of linear and nonlinear processing techniques as well as AI methods. The use of advanced processing techniques plays a key role in extracting meaningful information from biomedical signals. Among the most prominent methods are Fourier analysis, wavelet transform, and empirical modal decomposition. These approaches have been successfully applied to electrocardiogram (ECG) analysis, which enables the detection of cardiac abnormalities and the assessment of cardiovascular function with high accuracy and efficiency [1]; they have also been successfully applied to electromyography (EMG) analysis, which records the electrical activity of muscles [2]. Other common applications include electroencephalogram (EEG), which records the electrical activity of the brain and is fundamental in the diagnosis of neurological disorders such as epilepsy and sleep disorders [3]. These techniques allow complex signals to be broken down into simpler components, facilitating the identification of relevant biomarkers and subtle physiological abnormalities. These methods enable clinicians to gain a better understanding of disease pathology and tailor interventions to each patient's profile, thereby optimizing treatment outcomes. In parallel, the integration of AI methods has revolutionized the biomedical signal analysis landscape. Machine learning algorithms such as support vector machines, deep neural networks, and decision trees have demonstrated remarkable capabilities in pattern recognition and classification tasks [4, 5]. By iteratively learning from large data sets, these algorithms can identify intricate patterns within biomedical signals, outperforming traditional analytical approaches in terms of accuracy and efficiency. As a result, AI-based diagnostic systems are becoming promising tools for early disease detection, enabling early intervention and reducing the likelihood of chronic disease. The synergistic application of processing techniques and AI methods culminates in a paradigm shift towards personalized medicine. By harnessing the power of data-driven information, clinicians can develop tailored treatment regimens that take into account individual variations in disease manifestation and response to treatment.

2 Material and Methods

Data associated with biomedical signals and patient health status reports, in their natural form, present errors at the time of recording them that make it impossible to perform an accurate analysis, this is due to the fact that different people or systems are involved in this stage. Therefore, a stage of cleaning, unification, consolidation, and standardization of the data must be carried out so that they can be used and valuable information can be extracted. For this data treatment, there are a set of techniques: data cleaning, data wrangling, and record linkage.

The design of classification systems for medical applications demands meticulous attention to variables and data characteristics to ensure precise and insightful categorization of healthcare data. Within medical contexts, a plethora of variables come into play, including demographic data (such as age, gender, and ethnicity), clinical parameters (such as vital signs, laboratory results, and imaging findings), and diagnostic information (such as disease codes and pathology reports). These variables serve as the foundation for constructing classification systems tailored to specific medical tasks, whether it be disease diagnosis, patient risk stratification, or treatment outcome prediction. Despite the remarkable advancements facilitated by classification systems in medical applications, certain limitations persist.

Challenges related to data quality, including missing values, measurement errors, and data incompleteness, can hinder the performance of classification algorithms and undermine the reliability of predictions. Moreover, the interpretability of complex classification models, such as neural networks, poses a significant challenge in clinical settings, where transparent decision-making is paramount. Furthermore, issues pertaining to model generalization and scalability necessitate ongoing research efforts to develop robust and scalable classification frameworks that can effectively handle diverse and evolving healthcare datasets.

3 Sustainable Health

Sustainable health is a holistic approach to healthcare that aims to ensure the well-being of current and future generations while preserving natural resources and mitigating environmental degradation. It emerges at the intersection of public health, environmental sustainability, and social equity. In recent years, there has been an increasing recognition of the interdependence between human health and the health of the planet. This has led to efforts to incorporate sustainability principles into healthcare systems and practices. This shift in perspective involves rethinking healthcare delivery models to prioritize prevention, promote health equity, and reduce environmental impacts throughout the entire healthcare continuum. Sustainable health initiatives cover a wide range of strategies, from implementing green building standards in healthcare facilities to adopting renewable energy sources and sustainable procurement practices. Healthcare organizations globally are progressively adopting sustainability frameworks, such as the Sustainable Development Goals (SDGs) and the World Health Organization's (WHO) Health in All Policies approach, to direct their efforts towards achieving sustainable health outcomes.

In addition, digital health technologies and AI are transforming healthcare delivery by facilitating personalized and preventive care interventions, optimizing resource utilization, and reducing waste. However, despite the promising advancements in sustainable health, several limitations and challenges still exist. One of the main challenges is balancing the demands of economic growth, environmental conservation, and public health priorities within existing healthcare systems. Additionally, the unequal distribution of healthcare resources and environmental burdens disproportionately affects vulnerable populations, worsening health disparities and social inequities. The rapid pace of technological innovation in healthcare, including AI, raises ethical and regulatory concerns regarding data privacy, algorithm bias, and the potential for exacerbating health inequalities. Looking to the future, sustainable health projects are expected to tackle these challenges by using innovative solutions and interdisciplinary collaborations. This includes promoting sustainable food systems and active transportation, as well as advancing circular economy principles in healthcare supply chains. There is a growing emphasis on fostering partnerships between healthcare providers, policymakers, industry stakeholders, and community organizations to drive systemic change. Furthermore, the integration of AI and machine learning algorithms has the potential to improve the efficiency, effectiveness, and equity of healthcare delivery while supporting sustainable health goals. For example, AI-powered predictive analytics can identify population health trends, forecast disease outbreaks, and optimize resource allocation, enabling proactive interventions and informed decision-making.

4 Discussion

The integration of AI into medical care provides several advantages over traditional medical practice, particularly in terms of time efficiency and accuracy in diagnosing and treating patients. Unlike human doctors, who require years of training, AI systems can learn and adapt quickly, processing vast amounts of medical data in real time to generate precise diagnoses and recommend optimal treatments. Furthermore, AI systems' continuous adaptability enables them to stay current with the latest medical advancements, ultimately enhancing patient outcomes. This approach to precision medicine not only improves patient outcomes but also reduces the likelihood of adverse effects and therapeutic ineffectiveness [6, 7]. The integration of advanced technologies in biomedical signal analysis has implications for the overall goals of sustainability in healthcare [8, 9], responding to crucial clinical objectives.

First, the objective is to promote efficiency in the use of limited resources, both natural and financial, in a context where the demand for healthcare services continues to increase. This efficiency is essential to ensure the availability and accessibility of healthcare for all who need it. Clinical sustainability is aligned with ethical and moral principles that aim to preserve equity and fairness in healthcare access, it benefits patients and society and can lead to significant improvements in the quality and safety of patient care. Effective care focused on individual patient needs can be achieved by promoting innovation and optimizing clinical processes. The convergence of processing techniques and AI methods in biomedical signal analysis heralds a new era of precision medicine and sustainable healthcare [10]. By harnessing the power of data-driven information, healthcare systems can transcend traditional boundaries and deliver superior healthcare to patients. In a scenario where the demand for quality healthcare is constantly increasing, healthcare systems face challenges related to time constraints and excessive workloads. These challenges can compromise the quality of patient care. In this context, AI has emerged as a powerful tool, revolutionizing various aspects of patient care and medical research. The integration of AI in clinical medicine has the potential to enhance diagnostic accuracy and treatment outcomes, while also improving healthcare delivery efficiency, reducing costs, and enhancing patient experiences.

5 Conclusion

The integration of AI in healthcare presents ethical, legal, and safety challenges, such as ensuring data protection and fairness in its application. Addressing these challenges requires the development of stable algorithms, the integration of blockchain and big data technologies for secure data storage, and the establishment of independent oversight and ethics committees. However, it is important to maintain a healthy scepticism towards AI-generated information and to encourage public acceptance of these technologies in healthcare. Collaboration between humans and AI has the potential to improve patient outcomes. To optimize medical care, the integration of AI must be done in an ethical and responsible manner that promotes health equity.

References

1. M. F. Safdar, R. M. Nowak, and P. Pałka, Pre-Processing techniques and artificial intelligence algorithms for electrocardiogram (ECG) signals analysis: A comprehensive review, Computers in Biology and Medicine, vol. 170, p. 107908, Mar. 2024, https://doi.org/10.1016/j.compbiomed.2023.107908
2. M. K. Jalagam and V. K. Mittal, Studies on biomedical signal processing: A review, AIP Conference Proceedings, 2024, https://doi.org/10.1063/5.0159010.
3. A. Jain, R. Raja, S. Srivastava, P. C. Sharma, J. Gangrade, and M. R., Analysis of EEG signals and data acquisition methods: a review, Computer Methods in Biomechanics and Biomedical Engineering: Imaging & Visualization, pp. 1–26, Feb. 2024, https://doi.org/10.1080/21681163.2024.2304574
4. Rong, Guoguang, et al., Artificial intelligence in healthcare: review and prediction case studies, Engineering, vol. 6, no. 3, pp. 291–301, 2020, https://doi.org/10.1016/j.eng.2019.08.015
5. Kalaivani, K., et al., Prediction of biomedical signals using deep learning techniques, Journal of Intelligent & Fuzzy Systems, vol. 44, no. 6, pp. 9769–9782, Jun. 2023, https://doi.org/10.3233/jifs-230399
6. J. Feng et al., Clinical artificial intelligence quality improvement: towards continual monitoring and updating of AI algorithms in healthcare, npj Digital Medicine, vol. 5, no. 1, May 2022, https://doi.org/10.1038/s41746-022-00611-y
7. P. Sinonquel and R. Bisschops, Striving for quality improvement: can artificial intelligence help?, Best Practice & Research Clinical Gastroenterology, vol. 52–53, p. 101722, Jun. 2021, https://doi.org/10.1016/j.bpg.2020.101722
8. L. Liaropoulos and I. Goranitis, Health care financing and the sustainability of health systems, International Journal for Equity in Health, vol. 14, no. 1, Sep. 2015, https://doi.org/10.1186/s12939-015-0208-5
9. C. L. Ranabhat and M. Jakovljevic, Sustainable Health Care Provision Worldwide: Is There a Necessary Trade-Off between Cost and Quality?, Sustainability, vol. 15, no. 2, p. 1372, Jan. 2023, https://doi.org/10.3390/su15021372
10. G. Krishnan et al., Artificial intelligence in clinical medicine: catalyzing a sustainable global healthcare paradigm, Frontiers in Artificial Intelligence, vol. 6, Aug. 2023, https://doi.org/10.3389/frai.2023.1227091

Solar Desalination and the Application of the Life Cycle Analysis Method: A Mini-Review

Brayan Eduardo Tarazona-Romero, Alexander Meneses-Jácome, Yecid Alfonso Maldonado-Muñoz, and Alvaro Campos-Celador

Highlights

- Life cycle assessment (LCA) is presented as a foundational tool to promote circular economy practices in drinking water treatment systems.
- The presence of a "circular economy"associated with LCA applications for the drinking water sector unfolds a niche theme of scientific interest.

1 Introduction

Population growth, demographic increase, the impact of climate change, pressure on water resources, and lack of adequate infrastructure worldwide create a challenging landscape for addressing global water security [1]. According to recent research, it is estimated that about a quarter of the world's population residing in

B. E. Tarazona-Romero (✉)
Unidades Tecnológicas de Santander UTS, Bucaramanga, Santander, Colombia

University of the Basque Country UPV/EHU, Bizkaia, Spain
e-mail: btarazona@correo.uts.edu.co

A. Meneses-Jácome
Circular Bioeconomy Research Center (ε-BiO): Biotechnology and Environment Group, Universidad Autónoma de Bucaramanga (UNAB), Bucaramanga, Colombia

Y. A. Maldonado-Muñoz
Grupo de Investigaciones en Energía, Recursos, Sostenibilidad—GIRES, Universidad Autónoma de Bucaramanga (UNAB), Bucaramanga, Colombia

A. Campos-Celador
University of the Basque Country UPV/EHU, Bizkaia, Spain

A. Meneses-Jácome et al. (eds.), *The Circular Bioeconomy in Industry*, Waste as a Resource, https://doi.org/10.1007/978-3-031-84359-4_21

major urban areas currently relies on water supply systems that are under threat due to water stress [2, 3]. In this context, desalination is a viable alternative, based on the removal of salts from salt water generating two by-products: fresh water and brine, but for coastal regions and some industries (e.g., oil fracking, meat and sausage processing, microalgae harvesting, etc.), solar desalination also represents a key set of technologies for circular economy and improved competitiveness [4, 5].

Desalination is divided into two systems: direct (a single unit) and indirect (two units; desalination plant and energy system). The latter system, in turn, is classified as thermal and membrane. Both systems have advantages and disadvantages, sharing high energy consumption [6]. Currently, solar systems are being integrated into desalination technologies, giving way to the concept of solar desalination technologies, aimed at a more sustainable process. On the other hand, some critics claim that the brine produced is highly polluting [6], however, systematic studies on the environmental impact are limited [7].

Although desalination represents a potential solution to the problem of global drinking water scarcity, the feasibility of these projects is often compromised by environmental assessments that restrict their implementation. Therefore, this study proposes a bibliometric analysis to highlight recent research on how life cycle analysis (LCA) techniques are being applied [8], with the objective of comprehensively understanding how LCA is being used in desalination systems. This includes identifying key contributors, such as authors, leading journals, countries, and academic institutions leading research in this area.

2 Material and Methods

Figure 1 represents the methodological flow diagram, divided into three phases: (1) initial parameters, (2) bibliometrics, and (3) analysis of results. The database used was Scopus and was compared with web of science (WOS), finding a negligible difference in the number of documents. The software selected was VOSviewer (open access), and the words were selected based on relevance to the topic of study. Four search equations were proposed, reaching a final number of 150 documents, which were then filtered using VOSviewer and analyzed in terms of publications, citations, years of publication, institutions, and authors. Finally, 34 papers were selected to analyze in detail and identify current trends in the application of life cycle analysis in solar desalination systems.

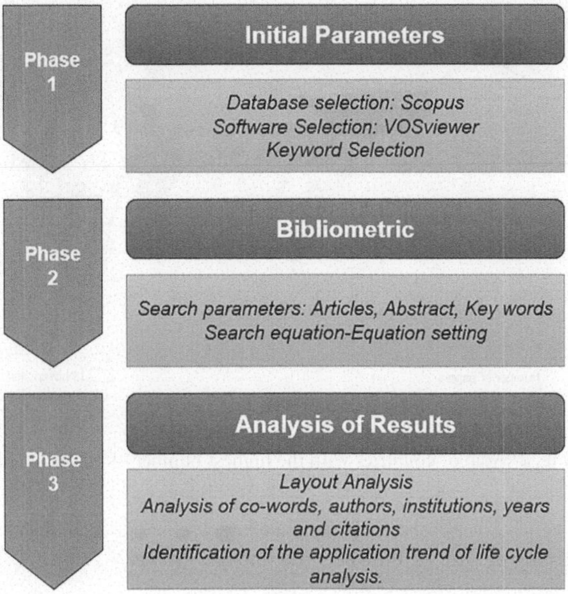

Fig. 1 Methodology flow diagram

3 Results and Discussion

In this section, the publications of documents categorized by the different countries will be analyzed, identifying the main producers of documents on life cycle analysis (LCA) in solar desalination systems. In this context, it is also intended to identify if there is a relationship between countries with low wealth indexes, countries with scarce water for human consumption (less than 5% of countries have improved sources of drinking water), and countries that carry out research on LCA in solar desalination systems.

Thus, there is a negative correlation between GDP per capita and inequality in access to water. In other words, limitations in access to drinking water are often accompanied by limited economic resources, which is intensified in impoverished regions of countries worldwide. This problem limits the ability of communities in these countries to equip themselves with desalination systems (see Fig. 2).

A total of 150 different papers (articles, reviews, book chapters, etc.) on LCA in desalination systems were evaluated during the research period between 2005 and 2023, identifying that 34 of all published papers emphasized solar energy. In this context, Fig. 3 graphically represents the year-to-year trend for the research period. It is highlighted that there is a varied growth with high peaks in publications during the last 5 years. The year with the highest incidence of publications so far for the research period is 2023 with 18 publications: 14 on LCA in desalination technologies and 4 on LCA in solar desalination technologies.

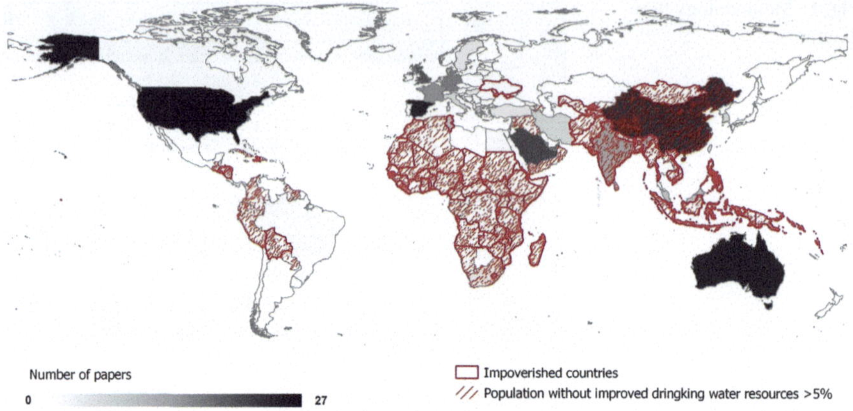

Fig. 2 Map of countries with the highest number of published papers on the research topic

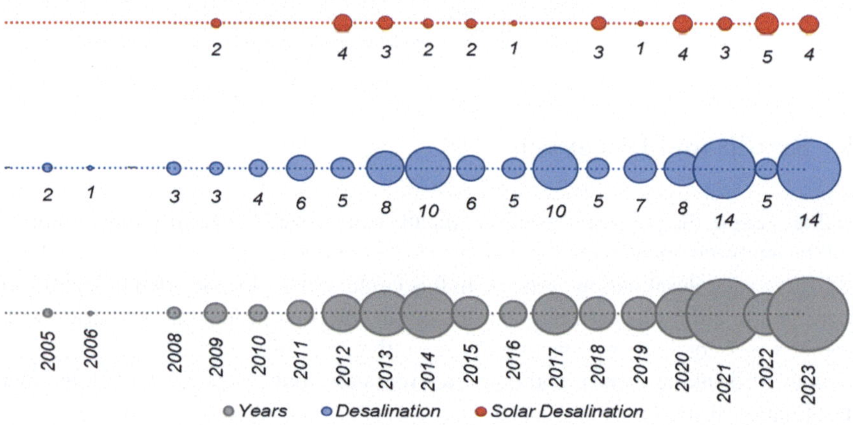

Fig. 3 Timeline of documents published in the period 2005 to 2023 for life cycle analysis in desalination and solar desalination

4 Conclusion

(1) The number of publications per year on LCA in solar desalination systems has increased in the last 3 years, showing a growing interest in the evaluation of the environmental impacts of the developed technologies. The difference in the number of papers published between countries is relatively low, so there is no evidence of net leadership by any continent in research topics in the area of study. (2) It also reveals that the researchers with the highest number of publications belong to the Asian continent. However, the institutions with the highest number of published papers and affiliated authors are not concentrated in a single continent. (3) The keyword co-occurrence analysis showed that most of the LCA publications on

desalination systems mainly dealt with water treatment, distillation, handling, production, and processing processes.

References

1. Tarazona-Romero BE, Campos-Celador A, Maldonado-Muñoz YA (2022) Energy Res Soc Sci 88:102510
2. McDonald RI et al (2014) Glob Environ Chang 27:96–105
3. Garcia C et al (2023) Sustain Cities Soc 96:104704
4. Xie W et al (2022) Chem Eng J 428(15):132624
5. Gao L, Zhang J, Liu G (2021) Desalination 2021(512):115148
6. Jones E et al (2019) Sci Total Environ 657:1343–1356
7. Panagopoulos A, Haralambous K-J, Loizidou M (2019) Sci Total Environ 693:133545
8. Lee K, Jepson W (2021) Desalination 509:115066

Environmental Peace and Bioeconomy: A Case Study of Initiatives in Ciudad Bolívar, Bogotá DC

Ramon Gabriel Aguilar Vega

Highlights

- Participatory community methodology IAP (Participatory Action Research) was implemented and tested as an effective approach to generate endogenous development of the bioeconomy based on natural capital knowledge.
- A multidimensional approach that considers interactions between the territory, society, and the environment is required to carry out impactful environmental education and promote effective and truly sustainable bioeconomy initiatives.

1 Introduction

This action-participation methodology combines theory with practice, promoting reflection and joint work to find alternatives that mitigate and repair existing environmental damage. Facing environmental realities, especially in vulnerable communities marked by violence, the destruction of strategic ecosystems, and a lack of community responsibility, is a challenge that requires commitment and action. This is why the students and teachers of the Jaba Kagüi hotbed have developed various activities such as workshops, talks, games, and outings, where they learn through serious and role-playing games. It is essential to address these problems

R. G. Aguilar Vega (✉)
Institución Universitaria Politécnico Grancolombiano, Bogotá, Colombia
e-mail: raguilar@poligran.edu.co

A. Meneses-Jácome et al. (eds.), *The Circular Bioeconomy in Industry*, Waste as a Resource, https://doi.org/10.1007/978-3-031-84359-4_22

comprehensively, involving all relevant actors in the search for sustainable and equitable solutions [1].

The inclusion of play and art in environmental education is crucial to promote comprehensive development in students since it allows them to connect emotionally and creatively with the subject, which increases their awareness and commitment to the preservation of the environment. It is important that educational systems encourage the integration of these methodologies in their study programs. Citizens must be responsible and committed to caring for our planet and stop so much greed toward natural resources. This lack of integration of play and art in environmental education limits the ability of students to understand the connective importance between art and play and to develop practical skills that allow them to apply their knowledge in everyday life. Furthermore, by not promoting art, the population would be limited in its ability to reflect and generate innovative ideas to solve daily environmental adversities.

For example, in terms of the goal of eradicating poverty, it is estimated that around 9% of the population of Bogotá lives in extreme poverty, which represents a significant challenge to achieving the goal of eradicating poverty in all its forms. Regarding the goal of ensuring quality education, the latest report from "Bogotá Como Vamos" shows significant progress in the capital. The literacy rate in the city is 97% [2]. indicating good access to education for the population [1]. However, there are still gaps in the quality of education, especially in rural and marginalized areas of the city. In relation to the goal of promoting gender equality, progress has been made in the inclusion of women in the political and economic life of the city. Currently, 34% of elected positions in Bogotá are held by women, reflecting a significant increase in female political participation. However, inequalities still persist in areas such as access to employment and the gender pay gap [3]. It can also contribute to the social inclusion of the most vulnerable communities, promoting equity and environmental justice. In many cases, marginalized or vulnerable communities are more threatened by environmental damage and strategic ecosystems. Education about climate change makes people more resilient and improves their ability to respond to disasters [4].

2 Material and Methods

For this project, a mixed methodology was developed, qualitative and quantitative, especially based on the IAP action methodology. This participatory community methodology IAP (Participatory Action Research) is an effective methodology where the focus is on the community and actions generate endogenous development for the bioeconomy. In this approach, the coexistence of two complementary processes is considered fundamental: on one hand, the process of knowing, which involves identifying and rigorously analyzing the reality in which the community is located; on the other hand, the process of acting, which involves the development and implementation of concrete strategies to transform that reality. The project also used social mapping. By mapping social networks, community organizations, available resources, and challenges facing a community, social cartographers can identify innovative and collaborative solutions [5].

The social diagnosis of the participating communities or the target population in vulnerable populations regarding environmental education was conducted. Surveys and interviews were carried out with different key actors in these communities, such as displaced persons, disabled individuals, reintegrated individuals, and former inmates. The aim was to identify the main needs and challenges they face in terms of environmental education [6]. Subsequently, educational workshops were designed and implemented to strengthen decision-making regarding their territory. Various teaching strategies, such as group dynamics, educational games, and informative talks, were used. Additionally, participants were provided with educational material, such as informative brochures, posters, and activity guides, to reinforce the knowledge acquired. Specific actions were analyzed regarding climate change for environmental conservation. Clean-up activities in green areas, tree planting, awareness talks on recycling, and waste reduction, among other activities, were carried out. The goal was to promote an environmental culture in these vulnerable populations within the social environment.

A monitoring of the implemented actions was carried out, evaluating the economic impact of these on the environmental awareness of the participants. Satisfaction surveys, individual interviews, and direct observation of the activities carried out were conducted. The level of acquired knowledge, changes in attitude toward the environment, and the willingness of the participants to continue promoting environmental education in their communities were measured [7]. The methodology used in this project combined qualitative and quantitative approaches in society. Innovative educational strategies were designed, and the impact of the implemented actions was evaluated [8].

3 Results and Discussion

The results are optimal in terms of education and the impact on the community of Ciudad Bolívar, children at an early age create environmental awareness and know the term bioeconomy to apply it in their reality. Waste that was previously seen as garbage is now viewed as an opportunity for work and employment. The fundamental concepts of the bioeconomy based on natural capital and their knowledge in the project acted as sustainable models. In Ciudad Bolívar, where workshops were held to manage waste, such as used oil, used oils were used to make biofuels and cleaning products, such as soaps and creams. The community develops a sustainable bio-based model, which implies that the basic materials for such an economy, clean chemicals and energy, are obtained from renewable biological products, including natural resources, as well as animal, plant, and waste sources.

The formulation of a sustainable model of integration with the community was fundamental in the project, this allowed the projection toward the sectoral organization for the preservation of the environment. In the games and roles, the concepts that suggest that humans have dominion over nature for instrumental reasons and that the role of environmental education is the acquisition of significant knowledge that is also considered as the beginning of environmental problems were discussed.

The historical reconstruction of memory was characterized by the idea that the human–nature relationship with the environment and ecosystem services, in its environmental dimension, has the function of instrumentalizing subjects for ecological and inclusive social practice. These results reflect the significant impact that environmental education can have on the formation of future generations committed to protecting the environment. Undoubtedly, continuing to promote this type of initiative is essential to build a more sustainable and respectful future for the planet [9].

Attendance at the workshops and field trips has been very impressive. We started with few people, but word of mouth has led the project and the foundation to bring in more than 40 people from all strata and with disabilities. Social inclusion is fundamental for any university career. What we were able to experience in the foundation is a neighborhood that has many social, economic, and environmental aspects that would be ideal to contribute to promoting principles based on justice and bioeconomy, to achieve complete merit and benefit all the people who live daily in this environment [10].

4 Conclusion

Through the implementation of a bioeconomy project, environmental reconciliation and peace can be promoted in the region by incentivizing collaboration among different local actors and promoting respect for the environment and life in all its forms.

The institutional capacity of local authorities can design and implement effective public policies in environmental matters. This includes institutions being able to carry out extramural projects and take knowledge to the most remote areas of cities, moving out of classrooms and improving concrete actions, including education. Educational games are effective tools for teaching environmental concepts and promoting the ancestral knowledge of grandparents in children, in terms of ecosystem conservation and bioremediation. Play, focused on games as a pedagogical tool, enhances cognitive, emotional, and social learning. Through educational games, children can experience, explore, and learn in a dynamic and fun way, acquiring knowledge, skills, and positive attitudes toward the environment.

To address these vulnerabilities in an integral manner, it is necessary to adopt a multidimensional approach that considers the full interactions between the territory, society, and the environment. This involves coordinated work between different actors and sectors, including local governments, community organizations, private sectors, and academia. It is fundamental to promote civic participation and the empowerment of communities so that we can actively contribute to the identification of risks, the formulation of policies, and the implementation of actions to reduce vulnerability. The construction of a culture of prevention and resilience in Ciudad Bolívar requires an effort together and compromise of all the actors involved.

References

1. Marcineková L et al (2023) Scand J For Res 38(4):265–274. https://doi.org/10.1080/02827581.2023.2211806
2. Onpraphai T et al (2021) Forests 12(12):1670. https://doi.org/10.3390/f12121670
3. Kalnbalkite A, Pubule J, Blumberga D (2022) Environ Clim Technol 26(1):75. https://doi.org/10.2478/rtuect-2022-0007
4. Corona Aguilar A, Gutiérrez Barbarrusa V (2018) Hacia La Inclusión Social Desde La IAP. Una Experiencia En Andalucía towards social inclusion from the IAP an experience in Andalusia. Empiria 2019(44):79–107. https://doi.org/10.5944/empiria.44.2019.25353. (In Spanish)
5. Alcaldía Mayor de Bogotá D.C. Plan de Ordenamiento Territorial Bogotá Reverdece 2022–2035. Documento Técnico de Soporte.2021. https://www.alcaldiabogota.gov.co/sisjur/adminverblobawa?tabla=T_NORMA_ARCHIVO&p_NORMFIL_ID=28934&f_NORMFIL_FILE=X&inputfileext=NORMFIL_FILENAME (In Spanish)
6. Rascón-Gómez MT, Fernández-Delgado FC, Alvarado-Jódar A (2019) Revista iberoamericana de educacion superior. Universidad Nacional Autonoma de Mexico, pp 157–169. https://doi.org/10.22201/iisue.20072872e.2019.28.434
7. Ministerio de Ciencia Tecnología e Innovación. Ciencia Y tecnología: fundamento de la bioeconomía; 2019. (In Spanish)
8. Rudenko S, Liashenko I, Komakha L, Hrebonozhko Y (2021) Bioeconomy education: implementation of the Eu bioeconomy strategy in Ukrainian higher education. Naukovyi Visnyk Natsionalnoho Hirnychoho Universytetu 202(5):165. https://doi.org/10.33271/nvngu/2021-5/165
9. Jiménez Mora J, Moreno Bayardo M, la Cruz D, Torres Frías J (2023) Significados Sobre Metodología de La Investigación En Programas de Doctorado En Educación. Una Exploración Desde Su Componente Curricular Educación 32(62):161–184. https://doi.org/10.18800/educacion.202301.007
10. Alcaldía Mayor B Informe de Calidad de Vida En Bogotá; 2022. www.bogotacomovamos.org. (In Spanish)

Index

A. Meneses-Jácome et al. (eds.), *The Circular Bioeconomy in Industry*, Waste as a Resource, https://doi.org/10.1007/978-3-031-84359-4